Planned Textbook by Automation Teaching Instruction Committee of Higher Vocational Education of Ministry of Education, PRC

Achievement in Development of Teaching Resources for Items of the National Vocational Students Skills Competition in 2008

Achievement in National Excellent Curriculum Development"Installation & Testing of Automatic Production Line"in 2010

Installation & Testing of Automatic Production Line

Lv Jingquan **Chief Editor**

Wang Jinfeng **Chief English Editor**

Li Jun, Tang Xiaohua, Zhang Wenming, Li Bo, Yao Ji **Editors**

Wang Jinfeng, Chen Zhifang, Xue Jian, Wang Huayi, Zhu Xiaoying, Ren Huifang, Zhao Wanhui, Li Meiying, Song Bo **Translators**

中国铁道出版社
CHINA RAILWAY PUBLISHING HOUSE

Introduction to the contents

The book consists of six parts. Chapter Zero is the project guidance, which mainly introduces guiding ideology and teaching design.Chapter One is the project start, which introduces national vocational skill competitions and typical Automatic Production Line(APL). Chapter Two is the project preparation, which comprehensively explains knowledge points, technical points and skill points that APL requires. Chapter Three is the project acceptance, which mainly introduces installation and testing process of the five working stations with typical APL as a carrier. Chapter Four is the project decision, which mainly introduces equipment installation, gas circuit connection, electric circuit design and connection in typical APL. Chapter Five is the project challenging, which briefly introduces trend in development of APL and application of modern technology. The disk includes audio materials of Chapter Zero, Chapter One, Vocabulary and Teaching PPT as well.

The book is available for a higher vocational education school curriculum textbook. It also can be used as the reference of the relevant engineers and technicians.

图书在版编目（CIP）数据

自动化生产线安装与调试 = Installation & Testing of Automatic Production Line : 英文 / 吕景泉主编 ; 王金凤译. -- 北京 : 中国铁道出版社, 2012.7（2018.6重印）

中国教育部高职高专自动化技术类专业教学指导委员会规划教材 2008年中国职业院校技能大赛赛项教学资源开发成果 2010年中国国家级《自动化生产线安装与调试》精品课程建设成果

ISBN 978-7-113-14056-4

Ⅰ. ①自… Ⅱ. ①吕… ②王… Ⅲ. ①自动生产线－安装－高等职业教育－教材－英文②自动生产线－调试－高等职业教育－教材－英文 Ⅳ. ①TP278

中国版本图书馆CIP数据核字(2012)第131854号

书　　名：Installation & Testing of Automatic Production Line
作　　者：Lv Jingquan　Chief Editor　Wang Jinfeng　Chief English Editor

策　　划：祁　云　　　　读者热线：（010）63550836
责任编辑：祁　云
编辑助理：绳　超
封面设计：刘　颖
责任印制：李　佳

出版发行：中国铁道出版社（100054，北京市西城区右安门西街8号）
网　　址：http://www.tdpress.com/51eds/
印　　刷：北京米开朗优威印刷有限责任公司
版　　次：2012年7月第1版　　2018年6月第2次印刷
开　　本：787 mm×1092 mm　1/16　印张：8.75　字数：193千
印　　数：3 001～4 000册
书　　号：ISBN 978-7-113-14056-4
定　　价：38.00元（附赠CD）

Brief Introduction to Authors and Translators

Lv Jingquan

Professor. Vice president of Tianjin Sino-German Vocational Technical College. He has obtained more than 20 vocational qualifications and certificates. He has studied at over 20 vocational education institutes and enterprise training centers in Germany, Singapore, Spain, Canada and Australia. He has published over 30 technical papers and 30 vocational education papers. Being a chief editor, he has compiled and published electro-mechanical textbooks and four planned textbooks of the National Tenth Five-Year Plan and the National "Eleventh Five-Year Plan". He hosted 6 national educational scientific research projects, and got 3 National Teaching Achievements, and organized and accomplished 14 National Excellent Curriculum. He has been in charge of regional integrated practice base construction supported by the Ministry of Education and the Ministry of Finance. He also in charge of the key research program of the Ministry of Education on "Construction and Practice of High Demand Technical Talents in Manufacturing Industry" and research program on "Teachers'Training Standard in Manufacturing" supported by The Ministry of Education and UNESCO.

Main Titles and honors:

- The Chairman of Automation Teaching Instruction Committee of Higher Vocational Education, Ministry of Education, PRC
- The 3rd National Distinguished Teacher
- Teaching team leader of state-level electro-mechanical specialty group
- The leader of National Excellent Curriculum "PLC Technology"
- The leader of National Excellent Curriculum "The Installation and Testing of Automatic Production Line"
- The leader of National Teaching Award of Core Technology Integration Construction Mode of Higher Vocational Electro-mechanical Major
- Member of Higher Vocational Talent Training Standard Evaluation Committee of the Ministry of Education
- Executive Member of Vocational Education Equipment Committee of China Academy for Vocational and Technical Education

Wang Jinfeng

Associate professor, senior interpreter. Director of Applied Foreign Languages Department of Tianjin Sino-German Vocational Technical College. She has 16 years' working experience in enterprise and has been an interpreter in industrial delegations to America, Canada, German, France, Italy, Holland, Luxembourg, Japan and Hong Kong of China. She has been engaged in vocational education for 12 years and published 6 papers. From the year 2005 to 2009 she has been trained in vocational education & teaching in Germany and America. She has been in charge of 4 science research programs and has been involved in the study & research on the theory & practice of modern vocational education.

Main Achievements:

- The Chairman of Higher Vocational Education Automation Technology Teaching Instruction Committee of the Ministry of Education
- Project Leader of Tianjin Excellent Curriculum, "Practical English"
- Project Leader of Tianjin Education Science "Eleventh Five Year's" Plan Program, "Research on Realization System of Vocational Education Modernization in China "
- Project Leader of the Teaching Reform Program of Tianjin Education Committee,"Research & Teaching Practice on Vocational Students' Professional English Skills Training Under the Situational Context."

Brief Introduction to Authors and Translators

Li Wen

Professor of Tianjin Sino-German Vocational Technical College. She has published over twenty papers, planned and published dozens of projected teaching materials. She has studied in Singapore and Hong Kong of China and acquired vocational qualifications and certificates. During her five years of enterprise career, she has been involved in dozens of technical development and reform projects and received a grant of patent. She has long been engaged in vocational education and made research in the field of vocational education theory and practice.

Main Titles and honors:

- leader of National Excellent Curriculum "Mechanical Design and Manufacturing Serial Courses"
- leader of National Excellent Curriculum "Mechanical Drawing and Measurement"
- Province-level College Distinguished Teacher
- leader of province-level teaching award Teaching Resource development and Application of Mechanical Design and Manufacturing Major
- The leader of province-level brand major "Mechanical Design and Manufacturing"

Li Jun

Professor. Director of Electrical Department of Harbin Vocational Technical College,. She has been engaged in teaching of automation for more than 20 years and taught over 10 related subjects. She has compiled 5 teaching materials and published 12 papers in EI periodicals and other related periodicals. She has been in charge of building up 15 training labs. She has been involved in 5 Harbin key research programs, such as "Campus Intelligent Electricity Saving Management System" and "College Graduates' Employer Evaluation".

Main Titles and honors:

- Member of Automation Teaching Instruction Committee of Higher Vocational Education, Ministry of Education, PRC
- Heilongjiang Province Distinguished Teacher
- Leader of National Excellent Curriculum "Machine Electrical Equipment and Upgrade"
- Leader of province-level Excellent Curriculum "Factory Electrical Control Equipment"
- Harbin Excellent Teacher

Tang Xiaohua

Deputy Dean of Teaching Affairs Office of Wuhan Electric Power Technical College, Associate professor/Engineer. He has been engaged in hydropower automatic product development in Wuhan Power & Scientific Development Co. Ltd and accomplished over 20 power enterprise renovations. Since 2004, he has been engaged in teaching and research on hydropower and electro—mechanical integration in the college. He has been in charge of one province Excellent Curriculum and awarded the first and second prize of China Electricity Council Teaching Award and the third prize of province scientific and technical development award. He has published over 10 papers in Chinese Core Journals, and compiled 4 teaching materials. He has also accomplished over 20 enterprise renovation programs.

Main Titles and honors:

- Hubei Excellent Youth Expert in 2006
- The 4th Technical Expert of Hubei Power Company
- The 4th Vocational Education Expert of Wuhan Electric Power Technical College
- Trainer of Hubei Power Company
- leader of National Excellent Curriculum "Automatic Operation and Monitoring of Generating Set in Hydropower Station"

Brief Introduction to Authors and Translators

Zhang Wenming

Dean of Changzhou Textile Vocational Technical College, associate professor, senior engineer. He is engaged in teaching and research in the field of electro-mechanical and experienced in design and development of carpet weaving machine and filling machine.

Main Titles and honors:

- leader of Specialty Development of Electro-mechanical Integration in Jiangsu.
- Leader of National Excellent Curriculum "Installation and Testing of Industrial Control System"
- Leader of Province Excellent Curriculum "Industrial Control Configuration and Touch Screen Technology"
- Leader of Province Excellent Curriculum "PLC Technology"
- Textbook of Configuration Software Control Technology has been awarded as Excellent TextBook in Jiangsu province.

Translator Team

The textbook is translated mainly by the English teachers from Tianjin Sino-German Vocational Technical College and in addition, translator (Zhang Yanlin) from Tianjin Iron & Steel Company. The translation tasks for chapters is as following:

Chapter Zero: WangJinfeng

Chapter One: WangJinfeng, Li Meiying

Chapter Two: Task One and Two: Ren Huifang

Task Three and Four: WangJinfeng, Zhu Xiao Ying

Task Five: Chen Zhifang

Task Six: Zhao Wanhui

Task Seven: Wang Huayi

Chapter Three: WangJinfeng, Ren Huifang, Li Meiying

Chapter Four: Xue Jian, Zhu Xiaoying

Chapter Five: Chen Zhifang, Wang Huayi, Zhao Wanhui, Song Bo

PREFACE 序

中国国家教育部与天津市人民政府联合国家有关部委，已经成功举办了 4 届全国职业院校技能大赛，技能大赛既是学生展示风采技艺比武的舞台，也是集中展示职业教育改革建设成果的平台；既是学校与企业合作沟通的渠道，也是社会各界及国（境）外同行了解中国职业教育的窗口。举办技能大赛的重要目的之一就是引领和促进职业教育教学改革和创新，赛项的开发与设计始终遵循教育与产业深度合作的原则，按照企业岗位要求和职业标准设计赛项、研制赛题，组织裁判工作和提供技术保障，因此，统筹安排赛项成果向教学资源的转化是各赛项执委会的责任之一。

"自动化生产线安装与调试"是教育部高职高专自动化技术类专业教学指导委员会牵头设计、2008 年首批入选的全国高职组比赛项目，比赛内容源于相关职业岗位具体要求，既考核了岗位通用技术和选手能力，又考察了选手团队合作精神、职业道德等综合素质。继 2008 年之后，高职高专自动化技术类专业教学指导委员会连续四年在全国举办该项赛事，使赛项设计日臻完善，影响日益扩大，2012 年又一次入选全国职业院校技能大赛高职组比赛项目。

《自动化生产线安装与调试》立体化教材，以赛项为依托，采取任务驱动形式，完整地体现了赛项对学生技术技能、职业素质、团队合作等方面的要求，实为学生日常实训、教师指导学生的重要参考。

该书发行 3 年来，再版 2 次，发行量逾 5 万册，还被译为英文版发行到东盟国家，对东盟技能大赛的赛项设计，产生了重要影响，同时，也是东盟国家学生 2012 年来华参与中国全国职业院校技能大赛的基础。

此次再版，作者进行了认真的修订，内容更加详实，进一步丰富了职业教育教学资源。

希望各赛项都能以技能大赛为平台，通过多种手段将大赛成果转化为教学资源，也希望作者再接再厉，通过组织、承办比赛，推出更多精品实训教材，更好地发挥技能大赛对教育教学改革的引领和指导作用，促进我国职业教育教学水平不断提高。

中国（全国）职业院校技能大赛执行委员会常务副主任
天津市教育委员会主任
靳润成
2012 年 6 月 6 日

PREFACE

In cooperation with the national ministries and commissions, the Ministry of Education, PRC and Tianjin Municipal Government have successfully held four sessions of the National Vocational Students Skills Competition. This competition is not only the stage for displaying the students' professional skills, but also the platform for showing the achievements of vocational education reform and development. It is not only the channel of communication between schools and enterprises, but also the window in which domestic and foreign professionals are able to see and learn about vocational education in China. One of the important purposes for holding the skills competition is to lead and promote the reform and innovation of vocational education. The development and design of the competition events pursue the principles of profound cooperation between education and industries. The competition events involving design, competition questions, referee task, and technical support is completed according to the enterprise' s requirements to the specific work position and its career standards. Therefore, one of the main duties of the executive committees of different competitions is to transfer competition achievements into teaching resource.

"Installation & Testing of Automatic Production Line" , which is designed by the Automation Teaching Instruction Committee of Higher Vocational Education, Ministry of Education, PRC , was first selected to be a competition event in 2008. The competition contents originate from the specific requirements of the relevant working positions. The competition not only assesses the competitors' general skills and ability, but also examines their overall quality of teamwork and professional ethics, etc. Since 2008, the Automation Teaching Instruction Committee of Higher Vocational Education has successively held the competition for four consecutive years, making the competition more perfect and having greater influence. "Automatic Production Line Installation and Testing" has once again become one of the competition events of National Vocational Students Skills Competition in 2012.

The comprehensive textbook of "Installation & Testing of Automatic Production Line" focuses on the competition events and task-oriented style, and reflects the competition requirements of the students' skills, professional quality and teamwork, etc. Therefore, it is an important reference both for students' daily training and for the teacher' s guidance for the students.

Since its publication three years ago, the textbook has been reprinted 2 times. More than 50,000 are in circulation. It has also been translated into English and issued to the ASEAN nations, which has great influence on the ASEAN Skills Competition Design. It is also the basic material for the ASEAN students who will come to China to participate in the National Vocational Students Skills Competition in 2012.

With this latest reprint, the author has conducted the revision. Therefore, vocational education teaching resource has been enriched.

It is hoped that achievements in competition events are to be transferred to teaching resources through various means regarding the skills competition as a platform. The author hopes to make further effort to release more quality training materials by way of organizing and holding competitions. The skills competition will play a leading and guiding role in education and teaching reform, thus improving national vocational education and teaching.

Executive Deputy Director of the Executive Committee
of National Vocational Students Skills Competition
Director of Tianjin Municipal Education Committee
Jin Runcheng
June 6, 2012

FOREWORD

This book is a comprehensive and practical English version of "Installation & Testing of the Automatic Production Line(2nd Edition)", the planned textbook by the Automation Teaching Instruction Committee of Higher Vocational Education of Ministry of Education, PRC. It focuses on work-processes, National Vocational Students'Skills Competitions, electro-mechanical skills training, and professional English teaching for electro-mechanical major students.

"Installation & Testing of the Automatic Production Line" is one of the competition events of the First National Vocational Students'Skill Competition held in Tianjin in June, 2008. This event was organized by the Ministry of Education of China, Tianjin Municipal Government, Ministry of Labor and Social Security, Ministry of personnel, Ministry of Construction, Ministry of Transportation, Ministry of Information and Industry and so on. During the 8th ASEAN Skills Competition held in Bangkok of Thailand in October, 2010, the Installation and Testing of Automatic Production Line was one of the competition events. The competition equipment, content and standards were provided from China. During the preparatory meeting for the 9th ASEAN Skills competition held in Jakarta in December, 2011, the YL-335B Automatic Production Line Equipment was designated as the competition equipment for the "Industrial Automation Competition Event". The 9th ASEAN Skills Competition will be held in Jakarta in October, 2012.

The ASEAN Skills Competition is under the World Skills Competition, demonstrating that fully recognized the YL-335B Automatic Production Line Equipment in 10th ASEAN nations. This also shows that the competition standards, contents and teaching resources of the Installation and Testing of Automatic Production Line of China Vocational Education have been internationally recognized.

The 5th National Vocational Students Skills Competition will be held in June, 2012, and the Installation and Testing of Automatic Production Line will be one of the competition events in the higher vocational group.

This book is work-process oriented, and it focuses on the core technologies of APL, which embodies the principles of excellence, sufficiency, and suitability in its use. It is a comprehensive textbook. Many teachers in vocational colleges have made favorable comments on the book, and it is widely used since its publication. The English edition of the "Installation & Testing of Automatic Production Line" aims at internationalize the competition resources and to serve the professional English courses as well as short-term electro-mechanical skills training.

The book is mainly translated by the teachers of Tianjin Sino-German Vocational Technical College and the engineers from enterprises. The teachers who are involved in the translation have received training for electro-mechanical and basic of the APL knowledge. In order to make the translation more professional and accurate, these teachers frequently communicated with electro-mechanical teachers during the translation of the book.

Task of the Editors

The book includes six main parts. Tasks have been distributed as follows: Prof. Lv Jingquan and Prof. Li Wen were in charge of project guidance Prof. Lv Jingquan and associate Prof. Tang Xiaohua were in charge of the project start; Associate Prof. Tang Xiaohua was in charge of the project preparation; Associate Prof. Zhang Wenming was in charge of the project acceptance; Prof. Li Jun was in charge of the project decision; Prof. Li Wen was in charge of the project challenging; Prof. Lv Jingquan and Prof. Li Wen were in charge of the project design and associate Prof. Yao Ji has aided and supported this book; Senior Engineer Zhang Tongsu has given guidance and provided various materials for the book, and also completed the mission manual and program list; Senior Engineer Li Bo has tested the program of the equipment and completed other various tasks. The whole book was planned and guided by Prof. Lv Jingquan.

Task of the Translators

The translation tasks have been distributed as follows: associate Prof. Wang Jinfeng was in charge of Chapter Zero. Associate Prof. Wang Jinfeng and Li Meiying were in charge of Chapter One. Ren Huifang was in charge of task One and two of Chapter Two; Associate Prof. Wang Jinfeng and Zhu Xiaoying were in charge of task Three and Four of Chapter Two; Associate Prof. Chen Zhifang was in charge of task Five of Chapter Two; Zhao Wanhui was in charge of task Six of Chapter Two; Wang Huayi was in charge of task Seven of Chapter Two. Associate Prof. Wang Jinfeng, Ren Huifang and Li Meiying were in charge of Chapter Three. Xue Jian and Zhu Xiaoying were in charge of Chapter Four. Chen Zhifang, Zhao Wanhui, Wang Huayi and Song Bo were in charge of Chapter Five. The translation team has obtained help from American NI Company, Chinese Yalong Science and Technology Group, Prof. Qian Yiqiu (Secretary-General of Automation Teaching Instruction Committee) and American teacher, Jason. The English version of the Installation & Testing of Automatic Production Line has been wholly edited by associate Prof. Wang Jinfeng.

Editors

June 6, 2012

CONTENTS

Chapter Zero Project Guidance—Teaching Design

Chapter One Project Start—Brief Introduction to Automatic Production Line

Chapter Two Project Preparation—Application of APL Core Technology

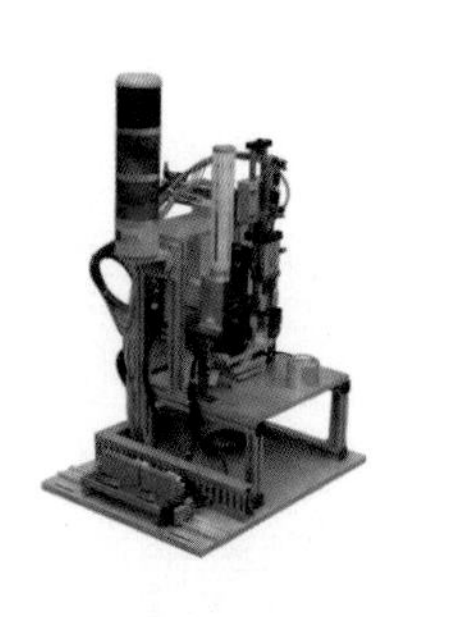

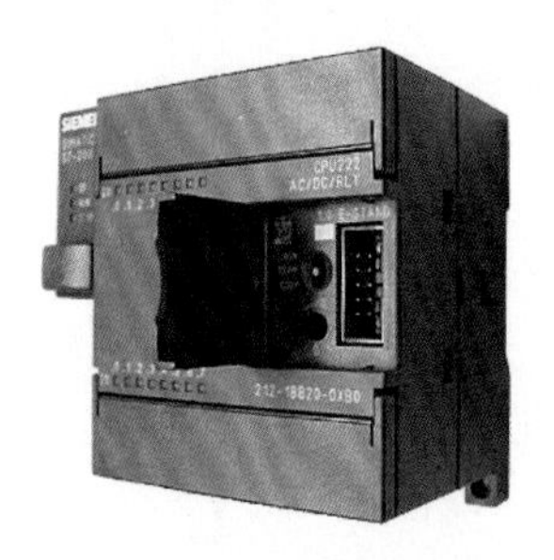

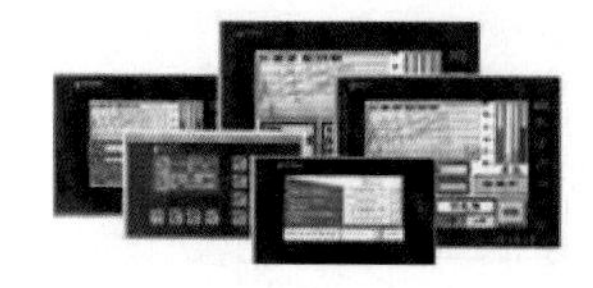

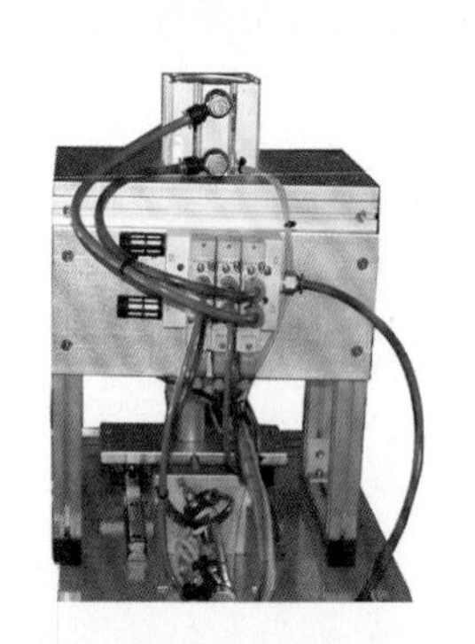

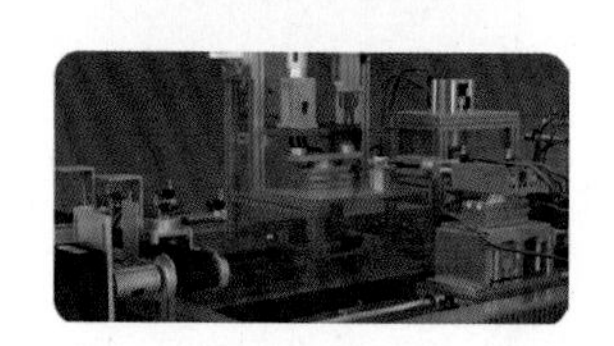

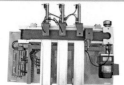

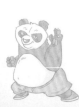

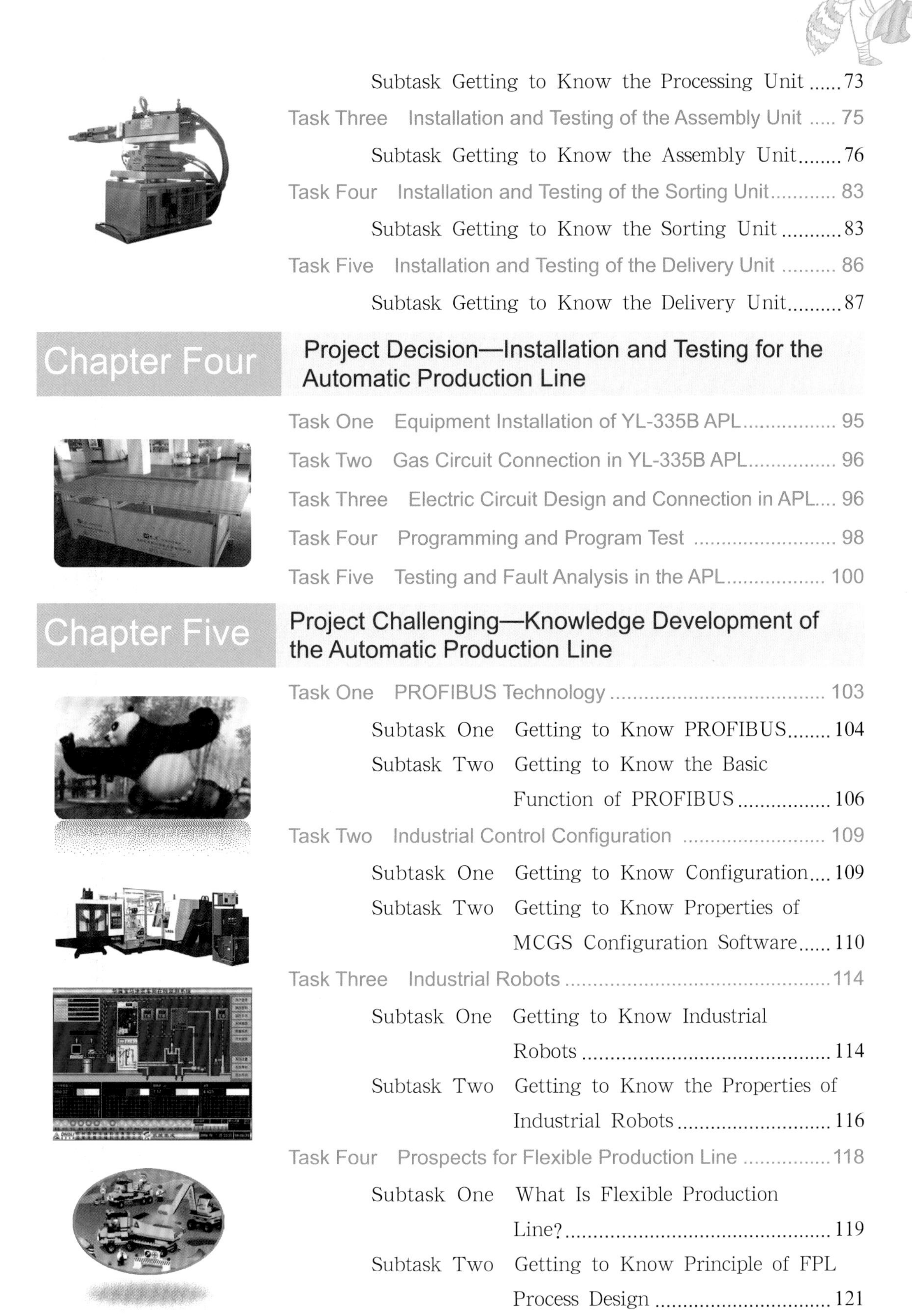

Installation & Testing of Automatic Production Line

Chapter Zero

Project Guidance —Teaching Design

Teaching through practice is considered as one of the most important ways and means for higher vocational students to obtain practical ability and multi-vocational abilities. It plays a very important role in the system of higher vocational education. Designing skills training practicum and professional skills training practicum and stimulating the students' interest in studying on their own are essential in training students to apply their acquired knowledge in production practice, which are the qualifications required in working post and also the prerequisite for obtaining ability for sustainable development.

Explanation One Guiding Ideology

To incorporate Integration of Program Core Technology into the curriculum development and teaching practice. To establish four integrations of core technology of professional courses on core knowledge and skills of the curriculum (See Fig. 0-1). Adapt to action guided teaching requirements, increasing the students' comprehensive adaptabilities to working posts. Train highly skilled personnel who need "short transitional period" or "no transitional period".

This project has won a second class medal for national teaching achievement in 2009.

Integration of program core technologies: Focus on professional training and clearly define a number of core technologies and skills. Plan the system of professional courses as a whole according to core technologies and skills. Make clear the core knowledge and skills in each course and to establish a teaching context (module) guided by working process. Incorporate theory, experiment, training, internships and employment. To build up an integrated and crossed teaching network of classroom, laboratory, attached workshop and production shop. Stressing theory and practice being paralleled, combined and cross-linked to each other longitudinally and transversely. The teaching process centers around core technologies and skills. Making the curriculum system and teaching contents serve the core technologies and skills and to enable the higher vocational graduates in this field to really acquire the ability to work. Aiming at training highly skilled personnel who need "short transitional period" or "no transitional period".

— quoted from "Study and Practice of Construction Mode of 'Integration of Core Technologies' in Higher Vocational Electro-mechanics" by professor Lv Jingquan.

This project has won a second class medal for the national teaching achievement in 2005.

Action Guided Teaching: From the point of view of teaching professional knowledge and skills, generally increase the students' comprehensive professional abilities. Enable them to systematically consider problems faced in their work, to understand the meaning of the work to be done, to be familiar with the work procedures and schedule, and to possess abilities to plan, perform and inspect on their own. As a prerequisite to be responsible for the society and to be able to effectively cooperate and communicate with other people. Work actively, carefully, on their own initiative, with a high sense of responsibility and quality; possess a sustainable ability in the associated technical field to adapt to the requirement in the future.

— quoted from "Application and Research on Action Guided Teaching Method in Higher Vocational Teaching Practice" by Professor Lv Jingquan.

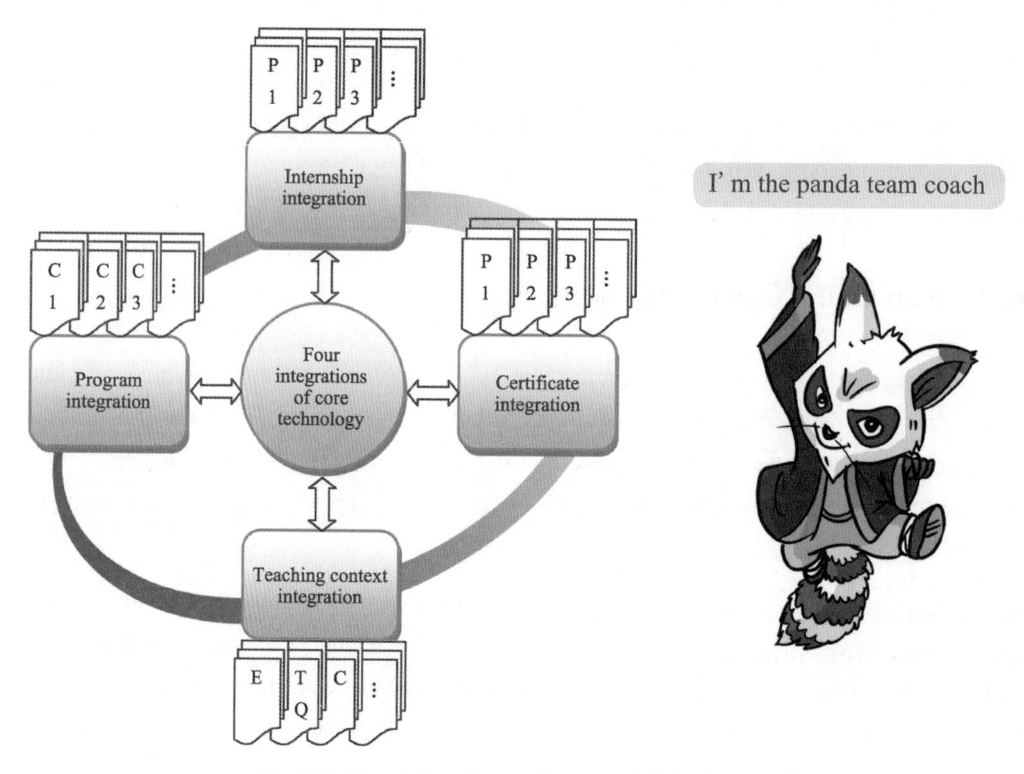

Fig.0-1 Four integrations of core technologies

Explanation Two　Teaching Design

Basic requirement: Equipment in production line for training and mechanical platform for typical Automatic Production Line (APL) should be provided. Each mechanism has a comprehensive function of electro-mechanical technology. The design concept of "Integration of core technologies" will be represented to build up a platform of action guided teaching mode.

Requirements for teachers: They should have a comprehensive knowledge of integration of electro-mechanical, be familiar with Automatic Production Line technology and possess great abilities in teaching and project developing.

Teaching carrier: Taking the YL-335 Automatic Production Line as the training platform to realize the design concept of curriculum development of core technology integration (see Fig. 0-2). Each project of the five sub-stations in the production line comprehensively covers the core technical skills in electro-mechanical field. It can be used to train and assess the students' ability of mastering core technology and application. It's an effective way to develop students' ability in technical innovation.

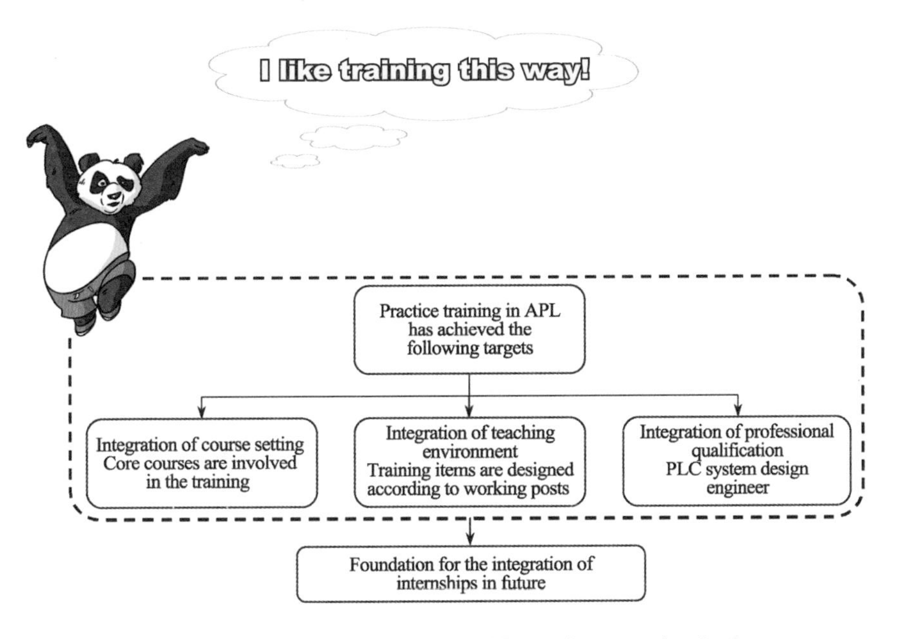

Fig.0-2 Relationship between APL and core technologies

Training mode: Groups of three students cooperate with each other in order to complete installation and testing of the five sub-stations in the automatic production line (see Fig. 0-3).

The general training equipment can be used for single station teaching, dual-stations teaching, multi-stations teaching and general on-line teaching. Each station covers different knowledge and skills. Schools or colleges may make related choices according to its specific requirements in vocational teachings.

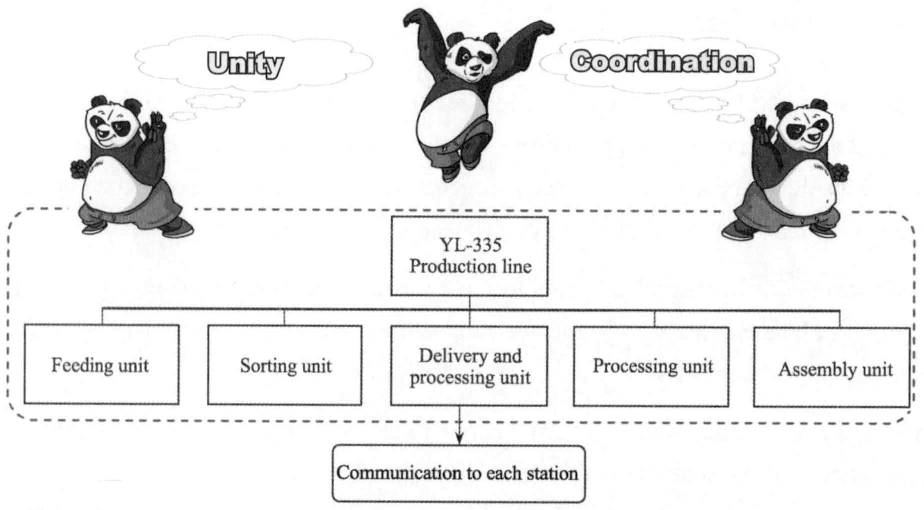

Fig.0-3 Functions of the production line

Training contents: The task in the project involves core technologies of mechanical and electrical engineering, mainly including: mechanical design, manufacturing process, mechanical assembly, installation of pneumatic parts, wiring of control circuit, installation of pneumatic solenoid valves and air pipe, PLC application and programming, application of frequency converter control technology, application of motion control technologies, electro-mechanical installation, connection, faulty diagnosis and testing.

Certification: The training contents include standard requirements for "PLC System Designers(level three)" and "Maintenance Electricians" to be qualified for certificates awarded by the Ministry of Labor and Social Security of the State.

Organizing the competition: Through the national higher vocational students skills competition, create a situation of "the higher education characterized by college entrance examination, while the vocational education characterized by skill contests". The abilities of the students in electro-mechanical of higher vocational colleges will be improved through contests of installation and test of Automatic Production Line.

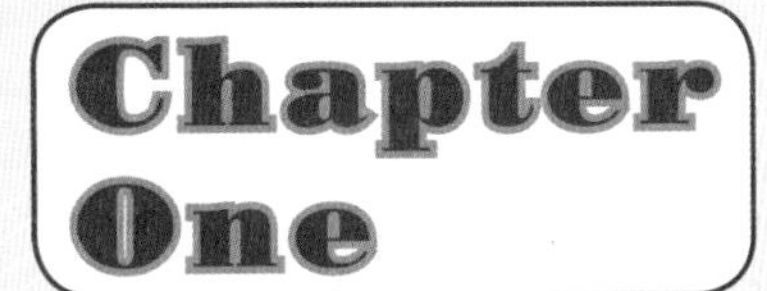

Project Start —Brief Introduction to Automatic Production Line

In the year of 2008, the Ministry of Education and 11 other ministries jointly held a National Vocational Students'Skills competition, in which installing and testing an automatic production line was an important event. The competition mainly comprised of two parts: installation of the Automatic Production Line and its running and testing. 72 contestants in 36 teams selected from all over the country participated in the competition. "Yalong Cup" 2009 National Higher Vocational Students' Skills Competition in Installing and Testing an Automatic Production Line was held. The all-around skills training in installing and testing the automatic line is playing an increasingly important role in the curriculum development in the field of electro-mechanical of higher vocational colleges. Fig.1-1 and Fig.1-2 respectively shows Opening ceremony and Teams from different cities.

Fig.1-1 Opening ceremony

Fig.1-2 Teams from different cities

Skills Competition

Firstly, the competition adheres to the combination of the competition with teaching reform, guiding the reform of teaching in the vocational education. Secondly, the combination of hi-tech (skills) with high efficiency, with enterprises (employers) taking part in the competition design and providing overall technical support and backing services. Thirdly, the combination of personal development with teamwork, emphasizing professional ethics and teamwork spirit while showing personal abilities.

Task Objectives

1. Getting to know the functions, roles and features of the Automatic production Line.
2. Briefing on the development of the Automatic Production Line
3. Getting to know the basic structure of YL-335 Automatic Production Line.

Task One Getting to Know APL and its Application

Fig.1-3 shows the APL of plastic case circuit breaker applied in ZhengTai Electric Appliance Co. Ltd. It includes automatic feeding, automatic riveting, five times power-on check, transient features' check, time delay features' check and automatic marking. It is controlled by PLC. Each unit has its own control and sounding and lighting alarm function. A perfect network system for the production line is built up by adopting a network technique, upgrading greatly the labor efficiency and product quality.

Fig.1-4 shows an automatic assembly line of automobile brake in an auto parts factory. Considering the factors of equipment performance, production cycling, general layout and material transportation, etc. It adopts standardized and modularized designs, various manipulators and programmable automation devices in order to achieve process automation in feeding, assembling, detecting, marking and packing. It also adopts network communication monitoring and data process realizes control and management.

Fig.1-3 APL of plastic case circuit breaker

Fig.1-4 APL of automobile brake

Fig.1-5 shows an automatic bottling line in a daily-use chemicals factory, which mainly performs production processes of feeding, can filling, sealing, detecting, marking, packing and stacking and fulfills the requirement for mass production.

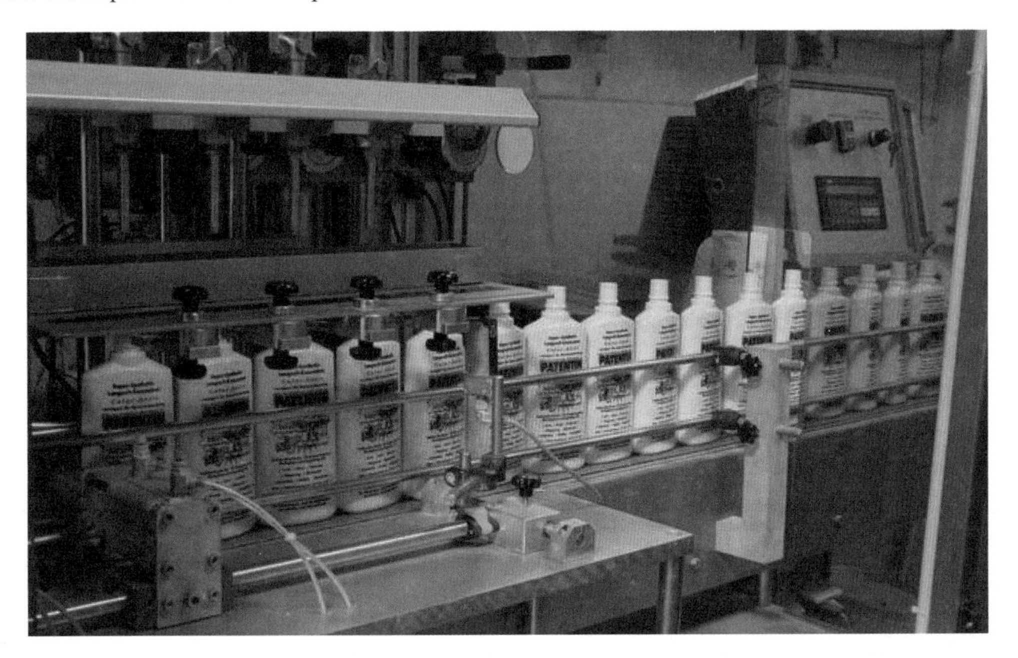

Fig.1-5 Automatic bottling line in a daily-use chemicals factory

In the disk provided, there are many other cases of production line. Take a look and think, what is an Automatic Production Line?

May I describe the APL in this way, Sir?

1. What is the APL?

The APL has gradually developed from a assembly line. It not only requires each mechanical unit in the line to automatically complete planned processes to make quality products, but also requires all the processes to be performed automatically in workpiece loading and unloading, positioning and clamping, workpiece delivering, workpiece selecting and even packing. It is designed to automatically operate in accordance with a specified program. This type of integrated electro-mechanical system working automatically is referred as Automatic Production Line (APL).

The task of APL is to achieve automatic production, but how?

To complete a predetermined processing task, the APL comprehensively combines and applies technologies of machinery, controlling, sensing, driving, internet and human-machine interface and integrates various machining devices in accordance with technological process with auxiliary devices and associates all the actions by controlling hydraulic pressure, pneumatic pressure and electrical systems.

2. Briefing on the development of the APL

As many fields of technologies are involved in APL, the development and perfection of APL technology has close relation with progress and interpenetration of various technologies. Therefore, the outline of the development of APL must be associated with the development of related technologies that entirely support the APL. The development of applied technologies are as follows:

Applied PLC

•It is a kind of industrial controller that mainly uses sequential control while making loop regulation auxiliary. It does not only have the functions of performing logic determination, timing, counting, memorizing and arithmetic operations, but also be able to control, on a large scale, switching and analog quantities, eliminating disadvantages of complex programming and non-standard external interfacing, insufficient use of machinery equipment presented while using industrial control computer in the switch control system, resulting in excess functions, higher manufacturing cost and poor adaptability to the project site. The PLC controllers have replaced many sequential controllers, such as relay control logic, for a lot of their advantages, and widely used in the control of automatic lines

Applied manipulator & robot

• Robot arms are widely used in automatic lines for work piece loading and unloading, positioning and clamping, transferring between procedures, removing surplus of material, processing and packing, etc. The third-generation intelligent robot currently under development will not only have the abilities of moving and operating, but also have distinguishing abilities in the sense of sight, hearing and touch. Automatic units that have the abilities of judgment, policy decision and languages are being more and more used in the automatic production lines.

Applied sensor

• With the development of material science and the advent of effects of solid physics in succession, the sensor technology, a complete and independent scientific system, has been formed and established. An applied "Intelligent Sensor" with micro-processor has come into being, which is designed to monitor various complex procedures automatically controlled in automatic production lines and is playing a very important role.

Applied hydrautic & pneumatic driving

• Especially in the case of pneumatic drive technology, which has won the general attention as it uses inexhaustible air as media, and is characterized by instant driving response, quick in motion, easier to make pneumatic parts, lower cost, convenient for centralized supply and long distance transportation. The pneumatic driving technology has become an independent technical field. It has been developing rapidly and get widely used in various industries, especially in automatic production lines.

Applied network

• Namely a leap in the applied network technology. Either a bus at the site or an industrial ether network enables each control unit in the automatic line to form a coordinated operating whole.

In general, the further progress of the integration of electro-mechanical technology that support the APL has made the functions of Automatic Production Line more complete, perfect and advanced, enabling it to perform technically more complex operation and produce products that require more advanced assembly process of the production line. As the information era has come, in view of a technology development, CIMS (computer integrated manufacturing system) will be a perfect condition for development of the Automatic Production Line.

Task Two Getting to Know YL-335 APL

Sir, I want to learn these skills and technologies!
If you want to give full play to them in future, you must start with simulation of YL-335 APL!

Note:

YL-335A was the contest equipment for the national vocational students' skills competition in APL Installation and Testing in the year 2008. It comprehensively uses many technologies such as pneumatic control technology, mechanical technology (mechanical transmission and mechanical connection), sensor application technology, PLC and networking, position control of stepping motor and frequency converter technology. It can simulate a control process very close to real situation , so that learners are able to enhance their skills in electro-mechanical integration.

YL-335B is a compatible product upgraded from YL-335A. It is the contest equipment for the "YaLong Cup" National Vocational Students' Skills Competition in APL Installation and Testing in 2009. Improvement of YL-335B has been made in extensibility, independency in single- station teaching, configuration flexibility and running reliability. It has included technical core contents of vocational education in electro-mechanical majors, which is good for teaching design and performance of the multi training curriculum in electro-mechanical major. YL-335B is a suitable supporting medium for the curriculum reform in electro-mechanical majors on the basis of working process and has been integrated with professional qualification standard of PLC "designer (level three)" made by the Ministry of Labor and Social Security.

1. Getting to Know the Basic Structure of YL-335B

YL-335B APL training equipment comprises of 5 units installed on aluminum alloy rail training platform: namely a feeding unit, a processing unit, an assembly unit, a delivery (transfer) unit and a sorting unit. At every station a PLC is installed to perform the function of control. Interconnection between PLC is made through RS-485 serial communications, which forms a distribution-type control system.

The task objectives of YL-335B APL is to transfer a workpiece from the feeding bin of the feeding unit to the processing table of the processing unit. After the piece is processed, it is transferred to the material table. Then the workpiece is embedded with small columns in different colors out of the feeding bin. The assembled products are then delivered to the sorting unit to be sorted and outputed. The sorting station sorts out the workpiece according to the material and the color of the workpiece. The diagram of YL335B is shown as Fig.1-6.

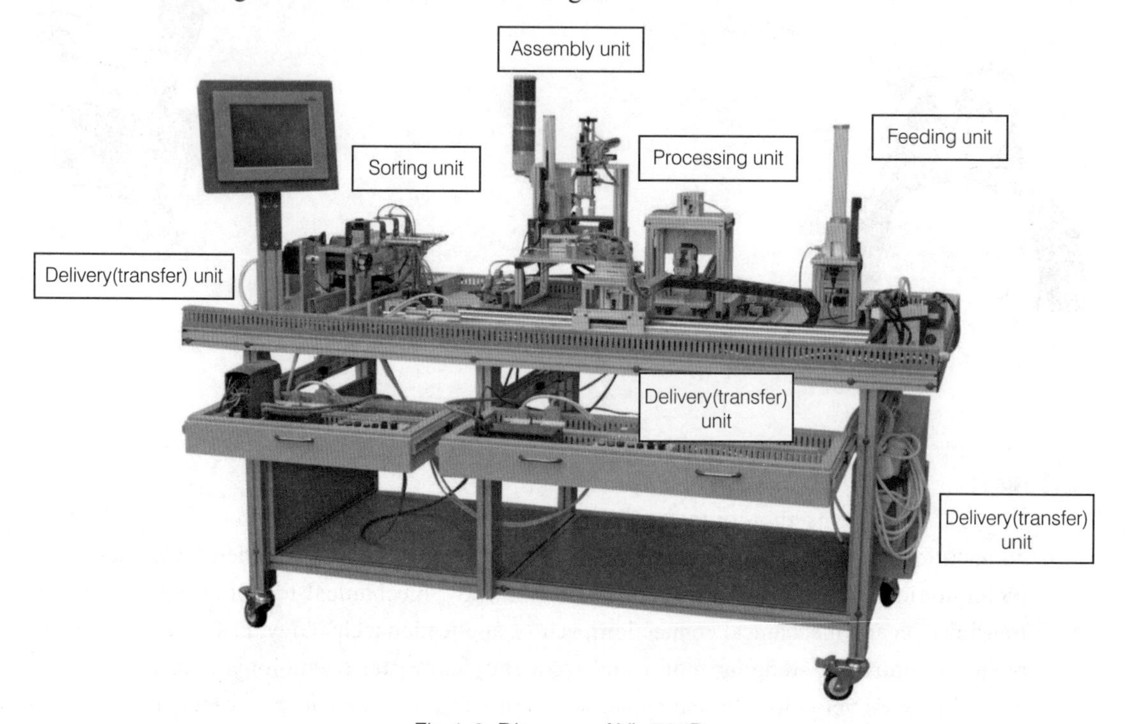

Fig.1-6 Diagram of YL-335B

Each unit works as an independent system and as an integrated electro-mechanical system as well. Basically, each unit has a pneumatic actuator. However, for the integral motion of the manipulator in the delivery unit, a servo-motor or a stepping motor is adopted to drive and perform precise position control. This driving system is characterized by its longer stroke and multi-positioning points and is a typical one dimensional positioning control system. The conveyor driving system of sorting unit uses an alternating transmission in which a three phase asynchronous motor is driven by the universal converter. The position control and converter technology is the most widely used electric control technology in modern industries and enterprises.

On YL-335B, there are many types of sensors installed to detect object movement positions, moving status, colors and characteristics.

Regarding the control, YL-335B adopts a PLC network control based on RS-485 serial communications. Each work unit is controlled by a PLC, and the distributed control mode of interconnection between PLC is realized through RS-485 serial communications. Users may select PLC from different manufacturers and RS-485 communications mode supported to establish a small

PLC network according to their needs. Mastering the PLC network technology based on RS-485 serial communications will lay a good foundation for further study of the site bus technology and industrial ether technology.

2. Getting to Know the Basic Structure and Function of the Feeding Unit

The feeding unit mainly includes workpiece storehouse, locking devices and pushing-out devices. Main devices include well-shaped workpiece storages, linear cylinders, photoelectric sensors and positioning devices, etc. The basic function of the feeding unit is automatically push out a workpiece to be processed from the feeding bin to the material platform in order to allow manipulator of the delivery unit to pick and take the workpiece to other units. The diagram of the feeding unit is shown in Fig.1-7.

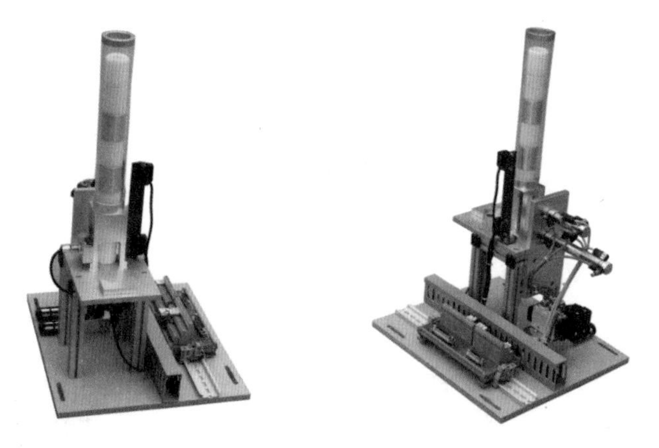

Fig.1-7 Exterior of feeding unit

3. Getting to Know the Basic Structure and Function of Delivery Unit

The delivery unit mainly includes linear moving devices and workpiece picking and delivering devices. Its main devices include driving motors, thin cylinders, pneumatic swinging tables, double guide bar cylinders, pneumatic fingers, limit switches and magnetic switches, etc.

Basic function of the delivery unit is to realize the accurate positioning of the material table to a designated unit. Grasp a workpiece and take it to a determined place and drop it. The diagram of the delivery unit is shown as Fig.1-8.

Fig.1-8 Exterior of the robot in delivery unit

4. Getting to Know the Basic Structure and Function of the Processing Unit

The processing unit mainly include workpiecemoving devices and workpiece processing devices. Its main devices include guide rails, linear cylinders, thin cylinders and workpiece gripping devices.

Basic function of the processing unit is to deliver a workpiece from the material table in the unit (which has been delivered by robot arm of the transfer unit) to the place under the pressing mechanism for once pressing. Then deliver the workpiece back to the material table, waiting to be gripped by the manipulator. The diagram of the processing unit is shown as Fig.1-9.

5. Getting to Know the Basic Structure and Function of the Assembly Unit

The assembly unit mainly includes is to assembl workpiece storages and workpiece carrying devices. Its main devices include workpiece storages, swinging tables, guide-bar cylinders, pneumatic fingers, linear cylinders and photoelectric sensors.

Basic function of assembly unit is to workpiece of small columns in black or white color from feeding bin into the processed workpiece. The diagram of the assembly unit is shown in Fig.1-10.

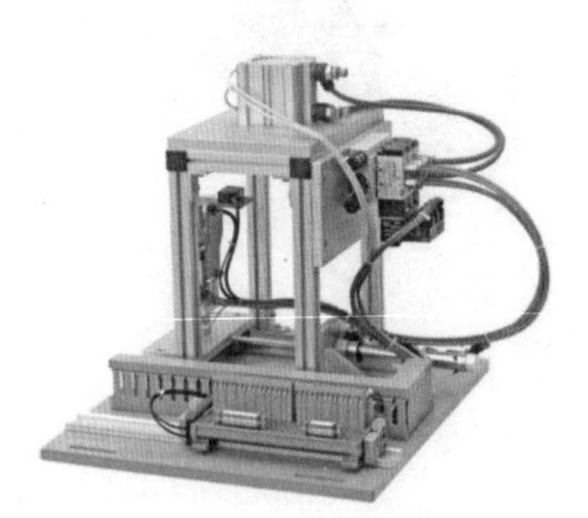

Fig.1-9 Exterior of processing unit

Fig.1-10 Exterior of assembly unit

6. Getting to Know the Basic Structure and Function of the Sorting Unit

The sorting unit mainly consists of a belt conveyer and a finished product sorting device. Its main devices include a linear belt conveyer, a linear cylinder, a three-phase asynchronous motor, a frequency converter, a photoelectric sensor and an optical fiber sensor.

Basic function of the sorting unit are Sorting a processed and assembled workpiece delivered from the previous unit and allow the workpieces to be carried away through its respective chute according to its color. The diagram of the pick-up unit is shown in Fig.1-11.

Fig.1-11 Exterior of Sorting unit

7. Control System of YL-335B

YL-335B adopts 5 PLC of Siemens S7200 series to respectively control 5 units as: delivery unit, feeding unit, processing unit, assembly unit, and sorting unit. PPI Serial bus is adopted for communications among them. Every working unit in YL-335B is controlled by PLC respectively. Each unit can work as an independent system. In addition, they can form a distribution type control system by interconnecting through the network.

When a working unit works as an individual system, its master instruction signals for equipment running and the running status display signal come from the push-button indicator light module of the unit, as shown in Fig.1-12. The ends of the indicator light and the button on the module are all connected to the terminal block.

YL-335B adopts MCGSTPC series touch screen as its human-machine interface (HMI). Master instruction signals for system operation (reset, startup, and stop) will be displayed through the touch screen HMI during the overall operation. In the meantime, various status information of the system operation will also appear on the HMI. The programming and using of the touch screen will be introduced in the following chapters.

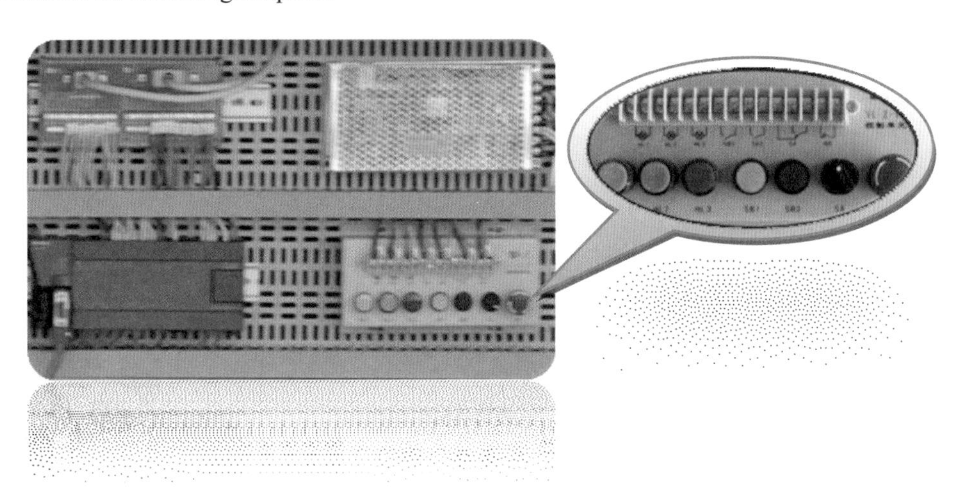

Fig.1-12 Unit control module

8. Power Supply

The external power supply to YL-335B is of three-phase and five-wire AC 380 V/220 V. Fig.1-13 shows the primary loop diagram of the power supply, in which the DZ47LE-32/C32 type three-phase and four-wire leakage switch is adopted for the main power switch. Each main load in the system is separately supplied with power through an automatic switch, in which the frequency converters are power supplied through DZ47C16/3P three-phrase automatic switches. DZ47C5/2P single phase automatic switches are used for PLC power supply in each working station. Besides, two DC24V6A voltage regulated power supplies of switch provided in the system are respectively used as direct current supplies to feeding, processing and sorting and delivering units.

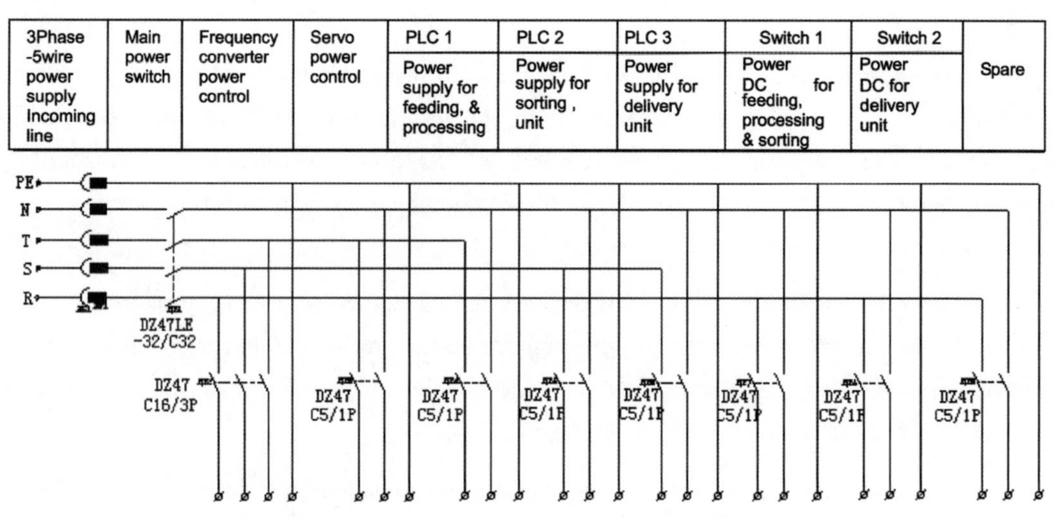

Fig.1-13 Diagram of primary loop for power supply block

9. Getting to Know the Characteristics, Parameters and Practical Training Projects of YL-335B

YL-335B is a set of semi-open type equipment. The structure of each working unit is characterized by relative separation of mechanical device from electrical control parts. The mechanical devices of every working unit are integrally installed on base plates. PLC devices which control the operation process for each working unit are installed on drawer plates on each side of the work table. To a certain degree, students may decide the number and type of units to form equipment as needed. An independent control system can be formed by 5 units at most or by 1 unit only. For a system comprising of more than 1 unit, the PLC network control program can represent the control characteristic of the APL.

YL-335B can perform so many trainings in technology freely, great! I want to learn it!

Main technical parameters of YL-335B:

① AC power supply: three-phase and five-wire AC (380 V±38 V) / (220 V±22 V) 50 Hz;

② Working temperature: −10～+40℃; ambient humidity: ≤ 90%(25℃);

③ Dimensions of the training table: length × width × height= 1920 mm×960 mm×840 mm;

④ Overall consumption: ≤1.5 kV·A;

⑤ Working pressure of gas supply: 0.6 Mbar (min,1 Mbar = 10^{11} Pa), 1 Mbar (max);

⑥ Safety protections: grounded protected, leakage protected and safe in accordance with related

GB standards. Use highly insulated safety sockets, high strength and safe cables and wires with insulated sleeve for experiment.

The mechanical parts of each unit are placed on the training tables for easy assembling or disassembling mechanical and pneumatic components, wiring for control circuit, installation of pneumatic solenoid valve and gas pipeline. Button/indicator light module, power module and PLC module of each unit are all installed on the drawer type module placing rack. Cable connection is adopted for the connection between modules and modules, modules and terminal plates on the training table. It meets the requirements of the comprehensive practical training.

YL-335B can be used for Trainings as follows:

① Practical training in automatic detection application;

② Practical training in pneumatic technology application;

③ Practical training in PLC programming;

④ Practical training in PLC networking configuration;

⑤ Practical training in electric control circuit;

⑥ Practical training in frequency converter application;

⑦ Practical training in motor driving and position control;

⑧ Teaching and practical training in automatic control technology;

⑨ Practical training in installing and testing mechanical systems;

⑩ Practical training in system maintenance and fault detection;

⑪ Practical training in configuration programming for touch screen.

I really want to participate in the competition, but what tasks to be done in it?

Main Tasks to Be Done in the Competition:

1. Equipment Installation

Complete assembling of parts of work units of feeding, processing, assembling, sorting and delivering units in YL-335B APL and install them on the worktable of YL-335B.

2. Pneumatic Circuit Connection

Connect pneumatic circuits according to requirements of the motion and control of pneumatic components based on the APL task.

3. Circuit Design and Connection

① Design electric (al) control circuit for the delivering unit according to control requirement, and connect the circuit according to the designed circuit diagram.

② Connect control circuits for feeding unit, processing unit and assembling unit according to a given I/O distribution table. For the sorting unit, design and connect the main circuit and control circuit for the frequency converter and connect the control circuit of the sorting unit according to the I/O terminals reserved for the frequency converter by the given I/O distribution table.

③ Connect communications network according to network control requirements of the APL.

4. Programming and Testing

① Making a control program for PLC and set parameters for the stepping motor driver and the frequency converter based on motion requirements in normal production and motion requirements in special cases.

② Testing mechanical parts, pneumatic components, detecting positions of components and making PLC control program to meet production and control requirements of the equipment.

Brief Summary

The modern automatic production equipment (Automatic Production Line) is most remarkably characterized by its comprehensiveness and systematicness. The comprehensiveness here refers to an organized combination of technologies in mechanics, electricity and electronics, sensors, PLC control, interfaces, driving, network communication and touch screen configuration programming, etc. And comprehensively applying them to production equipment; Systematicness refers to coordinative and ordered operation and organized combination of the mechanisms for detecting and sensing, delivering and processing, controlling, performing and driving under the control of PLC.

Installation & Testing of
Automatic Production Line

Project Preparation —Application of APL Core Technology

PLC is just like the human brain;
Photoelectric sensor is just like the human eyes;
Motor and conveyor belt is just like the human legs;
Solenoid valve set is just like the human muscles;
Human-machine interface is just like the human mouth;
Software is just like the central nervous;
Magnetic switch is just like the human sense of touch;
Linear cylinder is just like the human hands and arms;
Communication bus is just like the human nervous system;
Let's learn them together!

It is common that the PLC application technology, electrotechnics and electronic technology, sensor technology, intertace technology, network communications technology and configuration technology, etc.which are just like the human sensory system, motion system, brain and nervous system, are used in the automatic production line. In the following tasks, we will learn to use the core technologies mentioned above in YL-335B automatic production line See Fig.2-1. As a saying goes, "a workman must sharpen his tools if he is to do his work well."

Teaching concept of the project: "Learning by doing". Focuse on the core skills in the field of various technical applications in YL-335B .We'll explain the technical points of view through small tasks combing the professional qualification requirements of PLC system designer.

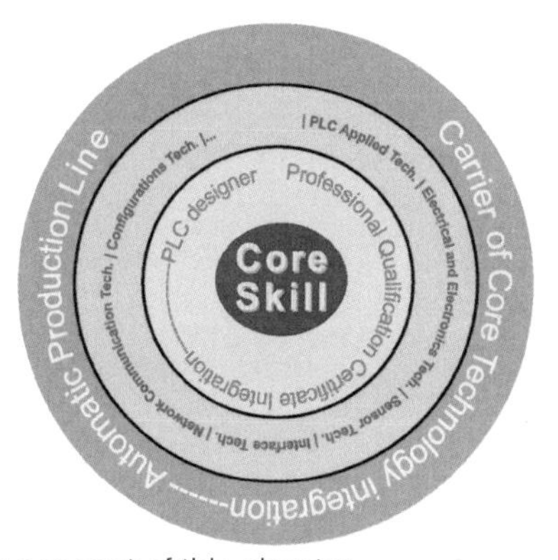

Fig.2-1 Teaching concept of this chapter

Task One Application of Sensors in the APL

Task Objectives:

1. Having a thorough understanding of the structure and characteristics of magnetic switch, photoelectric switch, fiber optic photoelectric proximity switch, inductive proximity switch and photoelectric encoder and characteristics of the electrical interface in the production line.

2. Be capable of conducting the installation and testing of various sensors in the Automatic Production Line (APL).

When the workpiece goes into the sorting station of the APL, it can be clearly observed by human eyes. However, how could the APL do it? How could we equip the APL with the function of human eyes?

Just like the human sense organs such as eyes, ears and nose, the sensor is the detecting component in the APL. It can feel the measured object and can convert it into the electrical signal and output it according to certain laws. In YL-335B Automatic Production Line, five types of sensors are mainly used including magnetic switch, inductive proximity sensor, photoelectric switch, fiber optic sensor and photoelectric encoder, as shown in Table 2-1.

Table 2-1 Sensors Used in YL-335B

Name of the sensors	Picture of the sensors	Graphics symbol	Use in YL-335B
Magnetic switch			Used for the cylinder piston's position detection of each unit in the APL
Photoelectric switch			Used for the workpiece detection in the sorting unit
Photoelectric switch			Used for the workpiece detection in the feeding unit
Fiber optic sensor		sharing the photoelectric switch's symbol	Used for the detection for the different color of the workpiece in the sorting unit
Inductive proximity switch			Used for the detection of the different metal workpiece in the sorting unit
Photoelectric encoder			Used for the driving belt's position control and revolution measuring in the sorting unit

Subtask One Magnetic Switch and Its Application

1. Brief Introduction of Magnetic Switch

In YL-335B Automatic Production Line, the magnetic switch is used for the position detection for all types of cylinders. As shown in Fig. 2-2, two magnetic switches are used to detect the positions where the cylinder of the mechanical hand extends and retracts.

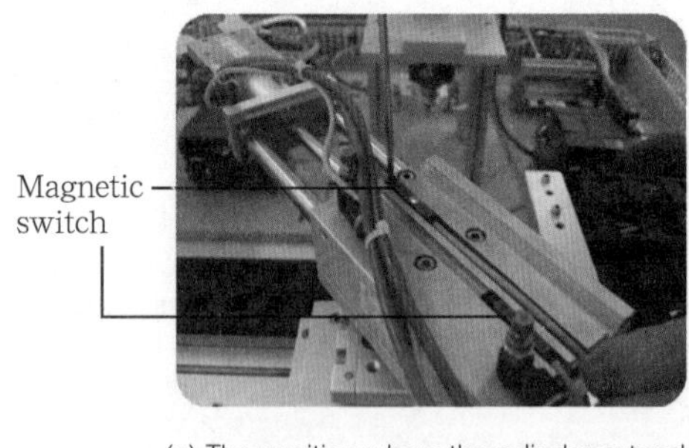

(a) The position where the cylinder extends

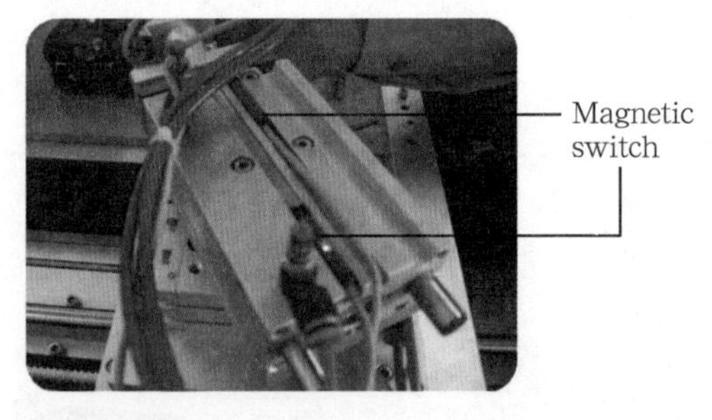

(b) The position where the cylinder retracts

Fig.2-2 Applications of magnetic switches

The magnetic proximity switch (magnetic switch for short) is a non-contact position detection switch, which enables the switch to respond at a high speed and unlikely abrade or damage the detected object. The magnetic switch is used to detect the presence of magnetic materials. Regarding the way of installation, there are wire-leading out type, plug-in components type and plug-in components relay type. According to the environment requirement of the installation site, either the shielded proximity switch or the non-shielded type can be chosen. Its real object picture and electrical symbol are shown in Fig.2-3.

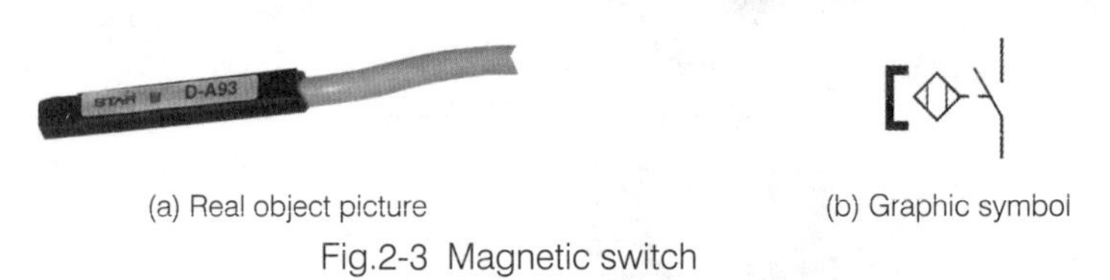

(a) Real object picture (b) Graphic symbol

Fig.2-3 Magnetic switch

When the magnetic substance is close to the magnetic switch sensor as shown in Fig.2-4, the sensor reacts and outputs switch signals. In practice, we install the magnetic substance on the measured object on the cylinder piston (or on the piston rod), and install a magnetic proximity switch at each end of the outer cylinder. The two extreme positions of the cylinder's movement can be identified respectively with the two sensors.

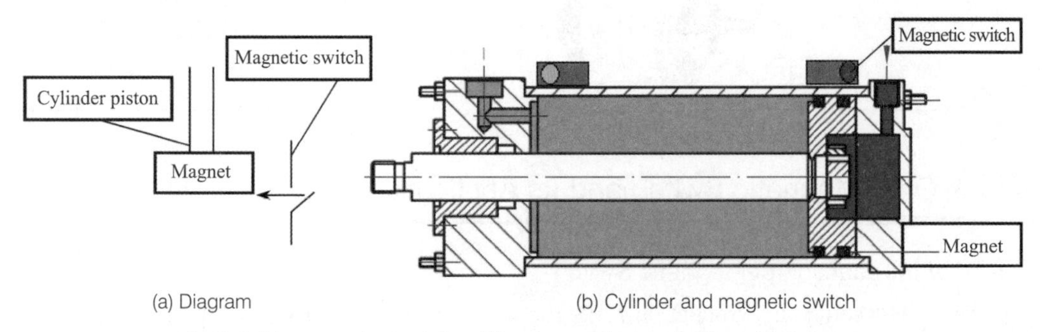

(a) Diagram (b) Cylinder and magnetic switch

Fig.2-4 Movement principle of the magnetic proximity switch sensor

As shown in the dotted box in Fig.2-5, the internal circuit of the magnetic switch employs common cathode connection, with the brown wire connecting PLC input and the blue line connecting the common terminal.

2 The Installation and Testing of the Magnetic Switch

In the control of the Automatic Production Line, the signal can be used to detect the cylinder's movement or location so that the workpiece can be detected whether it is pushed out or whether the cylinder is back.

(1) Electrical wiring and check

It is important to take into account the sensor's size, location, way of installation, wiring process, cable length, the surrounding work environment and other factors' influence upon sensors' operation. Now connect the magnetic switch with the PLC's input terminal referring to Fig.2-5.

There is one LED on the magnetic switch, which is used to display the sensor's signal state and to conduct the monitoring during testing and operation. When the cylinder piston gets close, the proximity switch will react and output signal "1" and LED is on. When the cylinder piston does not get close, the proximity switch will not react and output signal "0", and LED is off. See Fig.2-2 above.

(2) Installation and testing of the magnetic switch's on the cylinder

Magnetic switch is used together with the cylinder. If it is installed improperly, the cylinder's movement might be wrong. The magnetic switch does not have "sense" and act until the the cylinder piston moves towards it to a certain distance. This distance is normally called the "detected distance".

When install a magnetic switch on the cylinder, we should first install the magnetic switch on the cylinder, and the position of the magnetic switch is adjusted according to the requirement of the controlled object. It is quite easy to adjust it. Just fasten the bolts (or screw caps) with a screw driver when the magnetic switch arrives at the designated position as shown in Fig.2-6.

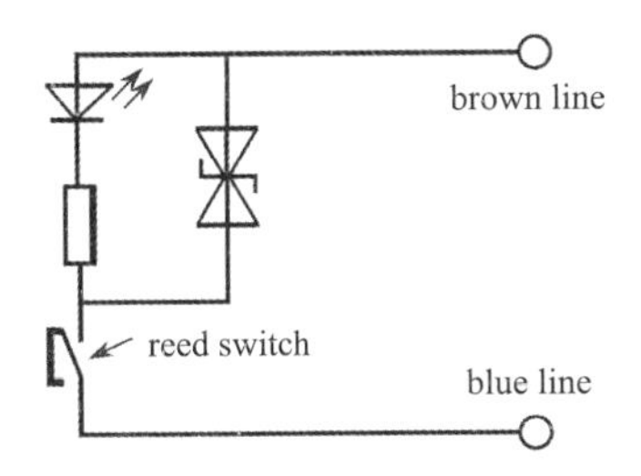

Fig.2-5 Internal circuit of the magnetic switch

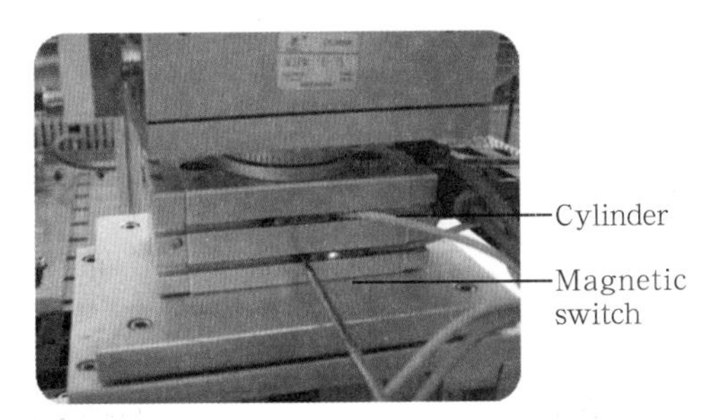

Fig.2-6 Magnetic switch adjustment

A magnetic switch is usually used to detect the cylinder piston's position. If other types of workpiece position will be detected, for example, a light-colored plastic workpiece, other kinds of proximity switch will be selected, eg. photoelectric switch.

Subtask Two Photoelectric Switch and Its Application

1. Brief Introduction to Photoelectric Switch

The photoelectric proximity switch (photoelectric switch for short) is usually used in relatively better and dustless environment. It has no influence on the detected object during its work. Therefore, it is widely used in the production line. In the feeding unit photoelectric switch is used to detect the workpiece in the feed bin as shown in Fig.2-7.

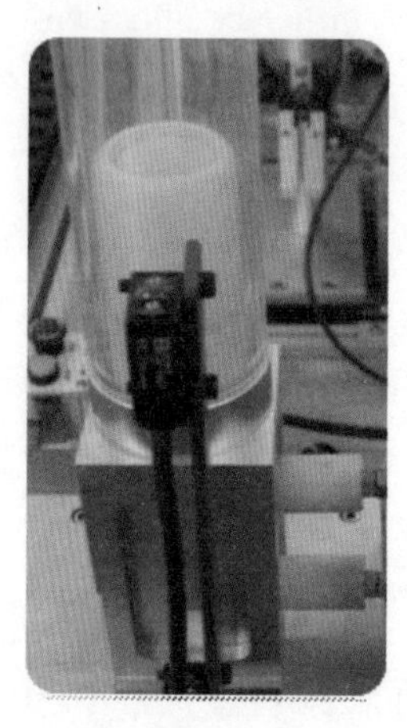

(a) There are workpieces in feed bin

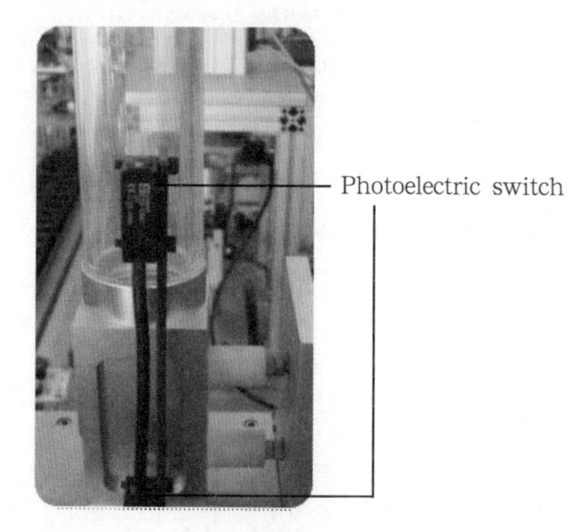

(b) There are no workpieces in feeding bin

Fig.2-7 Photoelectric switch's application in the feeding station

Two photoelectric switches, installed outside the feeding bin, are used to detect lack of workpiece and insufficient feeding. In case of this signal status of the two photoelectric switches can indicate whether there is feeding or whether the feeding is sufficient. In this unit, the small-beamed and the amplifier built-in diffused light photoelectric switch is used, whose shape and the adjustment knob together with display lamp on the top surface are shown in the Fig.2-8. The diffused light photoelectric proximity switch works by utilizing the reflected light from the detected workpiece. The light reflected from the workpiece is diffused light, so it is called the diffused light (ray) photoelectric switch. It consists of two parts—the light source (emitted light) and the photosensitive component (receiving light), with the optical transmitter and optical receiver on the same side.

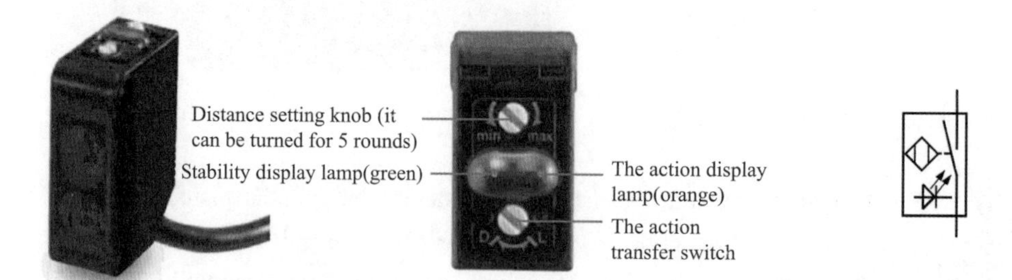

(a) Photoelectric switch's shape

(b) Photoelectric switch's graphic symbol

Fig.2-8 Photoelectric switch's shape, adjustment knob, display lamp and electrical symbol

During the operation, optical transmitting set keep emitting detection light (ray). If there is no object within a certain distance in the front of the switch, then no light is reflected to the receiver and the photoelectric switch is in the normal state without reaction. On the contrary, if there are objects within a certain distance in the front of the switch, as long as adequate light is reflected back, the receiver will get enough diffused light (ray) to make the proximity switch react and change the output state. Fig.2-9 shows the working principle diagram of the diffused light (ray) photoelectric switch.

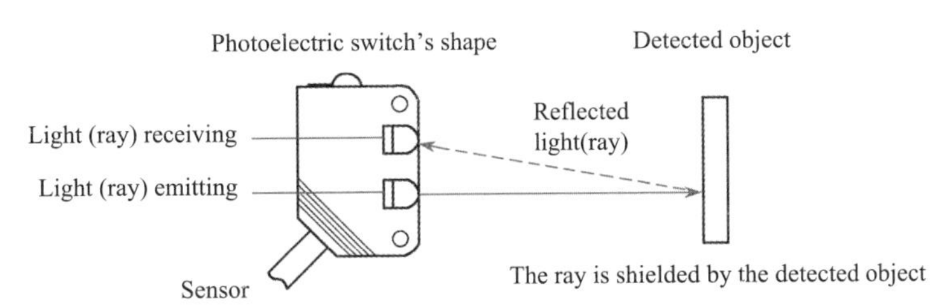

Fig.2-9 Working principle of the diffused ray photoelectric switch

2. The Photoelectric Switch's Application in the Sorting Unit

In the sorting unit of the Automatic Production Line, when the workpiece enters the sorting conveyor belt, the light(ray) emitted by the photoelectric switch in the sorting station will be reflected back to its own photosensitive component after meeting the workpiece, and then the photoelectric switch will output signal to start the operation of the conveyor belt.

(1) Electrical and mechanical installation

Install and fix the photoelectric switch according to the mechanical installation diagram and then perform electrical wiring.

Fig. 2-10 shows the schematic circuit diagram of the diffused light (ray) photoelectric switch used in YL-335B Automatic Production Line. In this diagram the photoelectric switch is equipped with the functions of power supply polarity and output reverse connection protection. Photoelectric switch with self-diagnostic function. When the after-setting environment changes (temperature, voltage, dust, etc.) meet margin requirements, the stability display lamp is on (if there is adequate redundancy, then the light will be on). When the light-receiving photosensitive component (element) receives the effective optical signal, the triode that controls the output will be turned on, and at the same time the motion display lamp will be on. This test can detect its optical axis deviation, the lens surface's (the sensor surface) pollution, the influence by the ground and the background, the external disturbances and the other sensor's abnormal phenomena and failures. It will be facilitate maintenance and steady operation. It also brings convenience to the installation and testing work.

Note: During the wiring of the sensor, attention should be given to the magnetic disturbance. Shine by sunlight or other light source should be avoided. Don't use in the site which producing corrosive gas, touches organic solvent and being much dust.

As shown in Fig. 2-10, the photoelectric switch's brown wire is connected to the "+" terminal of the PLC input module power, the blue wire is connected to the "-" terminal of the PLC input module power, and the black wire is connected to the PLC's input port.

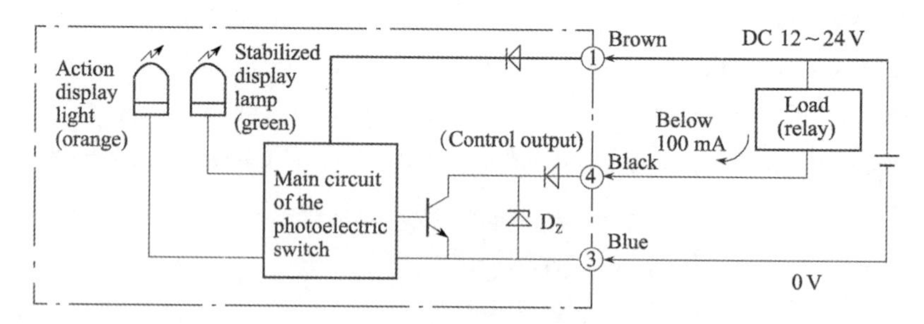

Fig.2-10 Circuit schematic diagram of the photoelectric switch

(2) Installation and testing

Photoelectric switch has such advantages as long detection distance, less limitation on the detected objects, fast response, high resolution ratio and easy adjustment. However, during the installation of photoelectric switch, we must ensure that the distance between the sensor and the detected object is within the "detection distance" range. At the same time, some other factors including the detected object's shape, size, surface roughness and moving speed should be taken into consideration. Testing process is shown in Fig. 2-11. In Fig. 2-11 (a), the adjusting position of the photoelectric switch is not in the proper place, so it is not sensitive to the workpiece and the display lamp is not on; In Fig. 2-11 (b), the photoelectric switch is in the right place and it is sensitive to the workpiece. So the display lamp is on and the stability lamp is on; In Fig. 2-11 (c), when no workpiece gets close to the photoelectric switch, the switch makes no output.

(a) Photoelectric switch is not installed properly.

(b) Photoelectric switch aas adjusted to the right place and detect the workpiece

(c) Photoelectric switch hasn't detected the workpiece

Fig.2-11 Testing of the photoelectric switch

Lock up the fixing nut after testing the photoelectric switch in the proper place.

The photoelectric switch's light source employs green light or blue light to distinguish the color. The photoelectric sensor sorts the products according to the different reflectivity of the surface color. In order to ensure the light's transmission efficiency and reduce the attenuation, in the sorting unit we use fiber optic photoelectric switch to identify workpieces black and white.

Subtask Three Brief Introduction to Fiber Optic Photoelectric Proximity Switch and Its Application

1. Brief Introduction to Fiber Optic Photoelectric Proximity Switch

Two fiber optic photoelectric proximity switches are installed respectively on the top of the conveyor belt in the sorting unit as shown in Fig. 2-12. A fiber optic photoelectric proximity switch consists of a fiber optic probe and a fiber optic amplifier and the two parts are separated. The end part of the fiber optic probe is separated into two optical fibers, which are inserted into the two fiber optic holes of the amplifier when used. The output terminal of the fiber optic photoelectric proximity switch is connected to PLC. To distinguish the white workpieces from the black ones, the sensitivity of the two fiber optic photoelectric proximity switches must be adjusted to different degrees.

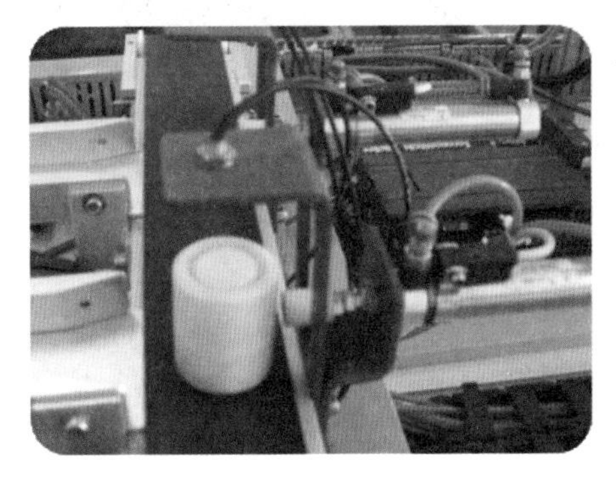

(a) Fiber optic probe

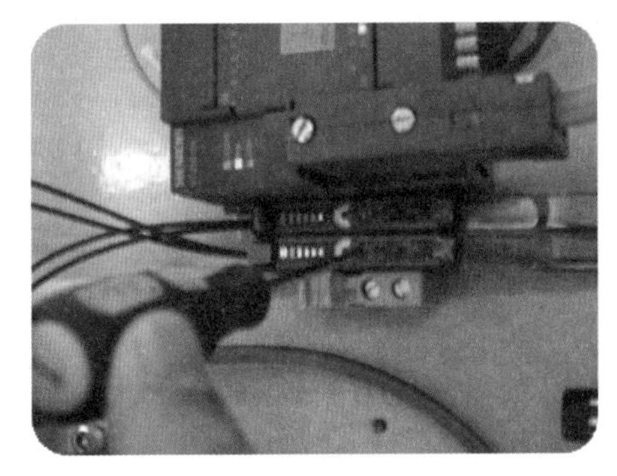

(b) Fiber optic amplifier

Fig.2-12 The fiber optic photoelectric proximity switch's application in the sorting unit

The fiber optic photoelectric proximity switch (the fiber optic photoelectric switch for short) is a kind of fiber optic sensors. In the sensing part of the fiber optic sensors there is no circuit connection, no heat production and just using a little optical energy, which makes the fiber optic sensor an ideal choice in the dangerous environment. Fiber optic sensors can also be used for the key production equipments' long-time reliable and highly stable monitoring. Compared with the traditional sensors, the fiber optic sensor has the following advantages of anti-magnetic interference, being able to work in harsh environment, long transmission distance and long service life. In addition, the optical probe has a relatively smaller size, so it can be installed in a very small space. The fiber optic amplifier should be placed where it is necessary. For example, fume or electric sparks can cause an explosion or a fire in some production process while optical energy can not be a fire source and it will not cause a disaster. Therefore, the fiber optic probe can be installed in hazardous locations and the amplifier unit could be placed in non-hazardous locations. The installation diagram is shown as Fig. 2-13.

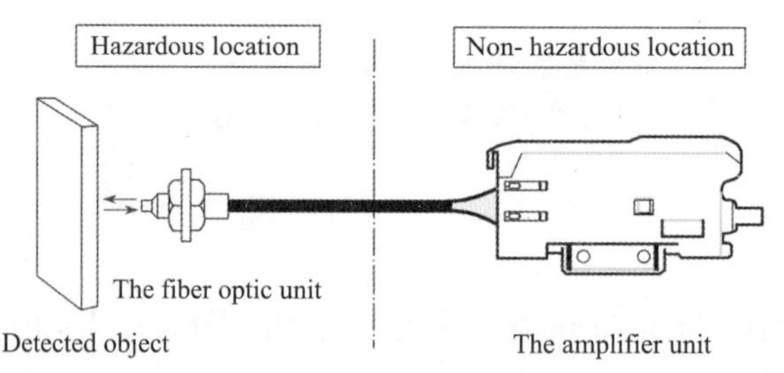

Fig.2-13 Installation diagram of the fiber optic sensor

A fiber optic sensor consists of a fiber optic probe and a fiber optic amplifier and the two parts are separated. According to the structure, fiber optic sensors can be divided into two categories: the sensing type and the light (ray) transmission type. The sensing type uses the optical fiber itself as the sensitive element so that the optical fiber can sense and transmit the detected information. The light (ray) transmission type puts the optical signal modulated by the detected object into the optical fiber and perform the measurement through the optical signal processing by the output terminal. The working principle of light (ray) transmission type is similar to the fiber optic sensor. The fiber optic photoelectric switch used in the sorting unit is the light (ray) transmission type, where the optical fiber is just used as a transmission route of the adjusted light (ray). The appearance is shown as Fig. 2-14, a light (ray)emitting terminal and a ray receiving terminal are connected to the fiber optic amplifier respectively.

Fig. 2-14 Fiber optic photoelectric switch

2. The Application of the Fiber Optic Photoelectric Switch in the Sorting Unit

Two fiber optic photoelectric switches are installed respectively on top of the conveyor belt in the sorting unit. The end part of the fiber optic probe is spearated into two optical fibers, which are separately inserted into the two fiber optic holes of the amplifier when used. The sensitivity of the fiber optic photoelectric switch's amplifier can be adjusted. When the sensitivity is too small, the fiber optic amplifier cannot receive the reflected signal from the black object which is less reflective. While the fiber optic amplifier's photoelectric detector can receive the reflected signal from the white object which is more reflective. Accordingly, we can adjust the sensitivity of fiber optic photoelectric switch to distinguish the white objects from the black ones, the two kinds of material, thus finishing the automatic sorting work.

(1) Electrical and mechanical installation

During the installation, first we should fix the fiber optic probe, install the fiber optic amplifier on the rail and then insert the two optical fibers of the end of the probe into the two fiber optic holes of the amplifier. Afterwards, we should go along with the electrical wiring according to Fig. 2-15. We must determine the power supply's polarity and the signal output line by the wire color.

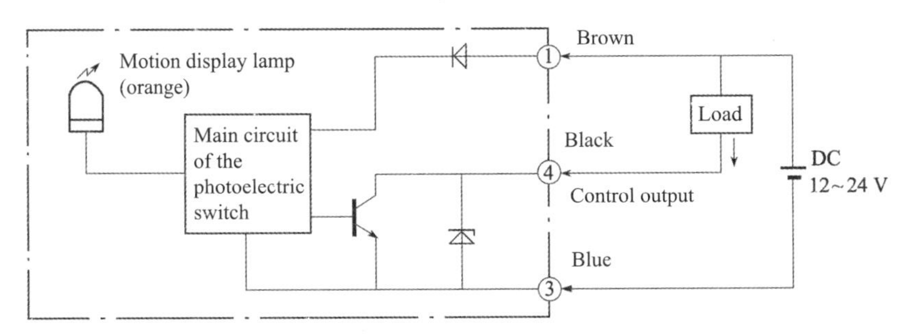

Fig.2-15 Circuit block diagram of the photoelectric sensor

(2) The sensitivity testing

How could we do the testing in the sorting unit? As shown in Fig.2-12 (b), the sensor's sensitivity can be adjusted by a screwdriver. Fig. 2-16 shows the top view of the fiber optic amplifier, in which we can see that the sensitivity can be adjusted by the high-speed sensitivity adjustment knob. During the adjustment, the "light (ray) input display lamp" changes. After fixing the detection distance, the "motion indicator" is on when the white workpiece appears below the fiber optic probe, representing that the workpiece is detected . When the black workpiece appears below the fiber optic probe and if the "motion indicator" is not on. Then the testing of the fiber optic photoelectric switch is completed.

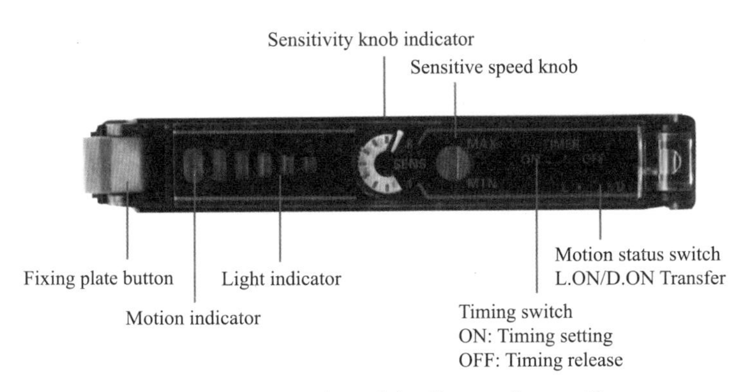

Fig.2-16 Top view of the fiber optic amplifier

The fiber optic photoelectric proximity switch becomes more and more popular in production lines. However, in some other environment, such as having more dust, being easily exposed to organic solvents and higher cost-effective requirement, we could actually choose some other sensors such as the capacitance proximity switch and the electric eddy current proximity switch substitute.

In the feeding unit, in order to test whether the workpiece to be processed is metallic, an inductive sensor is installed at the base side of the feeding tube, as shown in Fig. 2-17.

Fig. 2-17 Inductive sensor in the feeding unit

The electric eddy current inductive proximity switch is a kind of electric inductive sensors. It is a position sensor, which makes use of the eddy current effect. It consists of high-frequency LC oscillator and amplification processing circuit, which will generate electric eddy current inside the objects when metal objects get close to the oscillation induction head which can generate electromagnetic fields inside. The eddy current reacts on the proximity switch, cause the switch's oscillation capacity to fade and the inside circuit parameters to change, so to detect whether there is metal objects approaching, and control the switch on or off. This proximity switch can detect metal objects only. The working principle is shown as Fig.2-18.

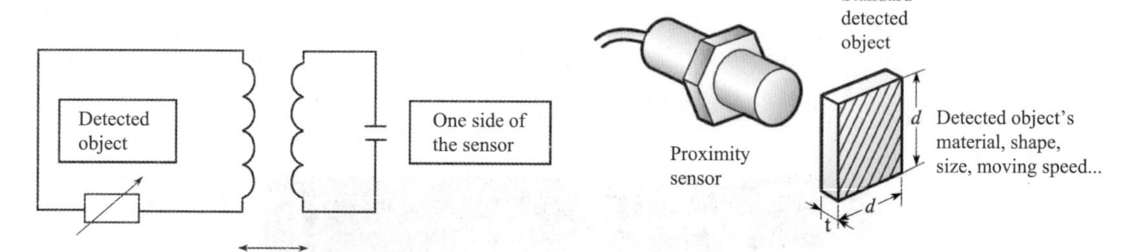

Fig.2-18 Working principle of the eddy current proximity switch

Fig.2-19 Proximity sensor and standard detected object

No matter which kind of proximity sensors are used, attention should be paid to the detected object's material, shape, size, movement speed and some other factors, as shown in Fig. 2-19.

In the sensor's installation and selection, we should seriously consider the detection distance and set distance to ensure the sensor's reliable movement in the production line. The installation distance is shown as Fig. 2-20.

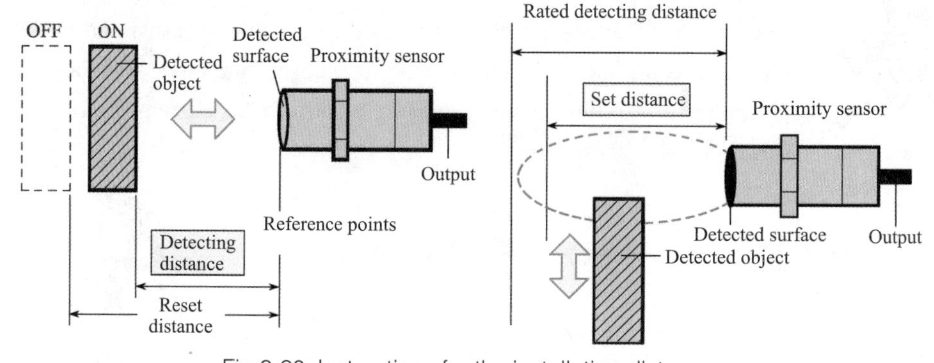

Fig.2-20 Instructions for the installation distance

When the situation does not require high accuracy, proximity switch can be used to count the products, measure the revolutions and even the angle of rotation displacement. While in some other situations when a high accuracy is required, the photoelectric encoder can be used to measure the rotation displacement or measure the linear displacement.

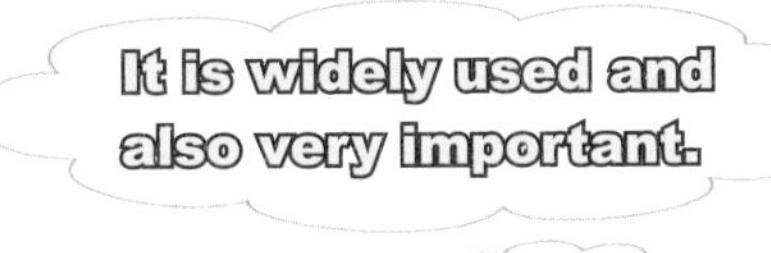

Subtask Five Brief Introduction to Photoelectric Encoder and Its Application

Of the controls in the sorting unit in YL-335B Automatic Production Line, the conveyor belt's positioning control is carried out by the photoelectric encoder which is also supposed to measure the motor's rotation rate. Fig. 2-21 shows two applications of the photoelectric encoder in the sorting unit.

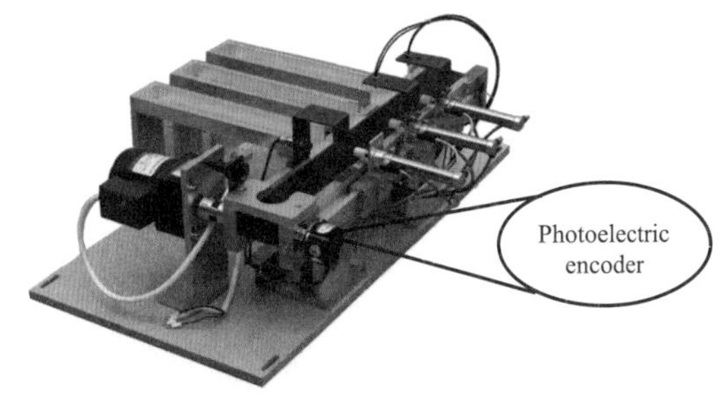

Fig.2-21 Application of the photoelectric encoder in the sorting unit

The photoelectric encoder is a kind of sensor that transfers the displacement magnitude of mechanics and geometric into the pulse or digital magnitude through photoelectric conversion. It is mainly used for the speed or position (angle) detection. The typical photoelectric encoder is composed of disk, detecting mask (grating), photoelectric conversion circuit (including light source, photosensitive element, signal conversion circuit) and mechanical components. Generally speaking, according to pulse generation ways, the photoelectric encoder can be divided into three categories: the incremental encoder, the absolute encoder and the compound encoder. The incremental photoelectric encoder is commonly used in the production lines. Its structure is shown as Fig.2-22.

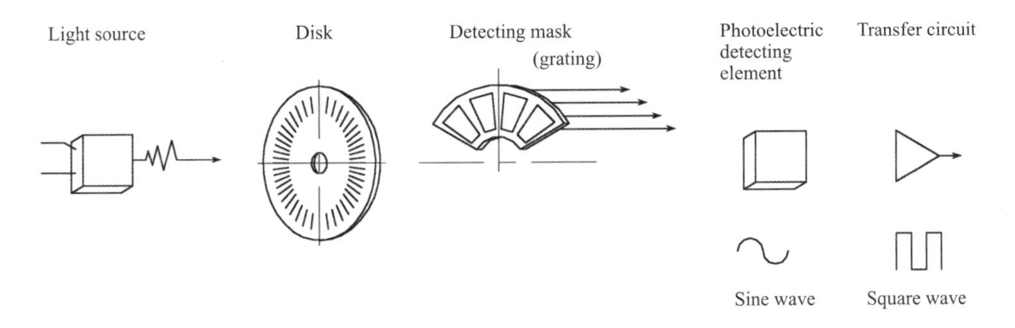

Fig.2-22 Structure of the incremental photoelectric encoder

The disk's fringe number of the photoelectric encoder determines the sensors' the least resolution angle: the resolution angle α=360°/fringe number. If the fringe number is 500, the resolution angle α = 360°/500 = 0.72°. There are two groups of fringes on the mask of the photoelectric encoder: group A and B. The fringe A and B are staggered by 1/4 pitch. Signals generated by photosensitive element corresponding to the the two groups of fringes deviate by 90° , which is used to judge direction. Furthermore, there is a transparent fringe Z in the inside circle of the photoelectric encoders'disk , which is used to generate a pulse per revolution. This pulse becomes a transfer signal or a zero sign pulse. Its output waveform is shown as Fig.2-23.

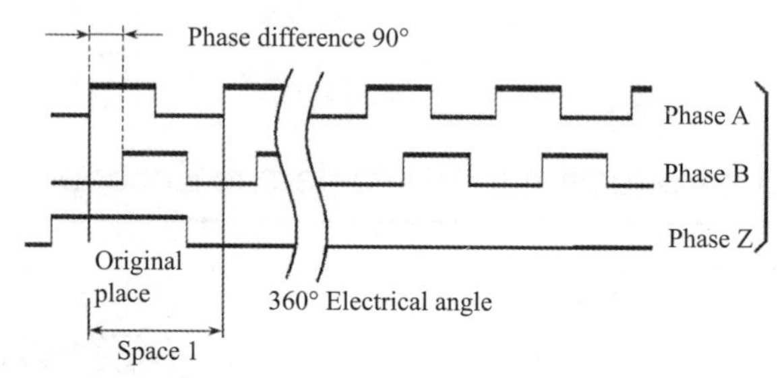

Fig.2-23 Diagram of the incremental encoder's output

A and B two-phase rotary encoder with a 90° phase difference is used in the YL-335B sorting unit to calculate the workpiece's position on the conveyor belt. The encoder is directly connected to the conveyor belt's drive axle. This rotary encoder's three-phase pulse uses NPN collector open circuit output , with the resolution of 500 lines and working power of DC 12~24 V. Z-phase pulse is not used in this unit and the A and B two-phase output terminal is directly connected to the PLC high speed counter's input terminal.

When we calculate the workpiece's position on the conveyor belt, we should make sure about the distance between every two pulses, that is, the pulse equivalency. In the sorting unit, the drive axle's diameter is d=43 mm. So when the geared motor rotates a revolution, the workpiece's movement distance on the belt, $L = \pi d = 3.14 \times 43$ mm = 135.02 mm. Accordingly, the pulse equivalency is μ= L/500 = 0.27 mm.

When the workpiece moves from the centerline of unloading to a point that is 164 mm away from the center of the first pusher, the rotary encoder sends out 607 pulses.

In production line there are many other advanced sensors, such as the CCD (Charge Coupled Device) used in the product quality testing, the mask sensors and the magnetic grid sensors used in the linear displacement detection, and some other sensors. We can choose the suitable sensor according to the requirement of the Automatic Production Line.

Summary of Knowledge and Skills

A wide range of sensors are used in various Automatic Production Lines. In most cases, the reason of the Automatic Production Line does not work properly is that because the sensor's

installation and testing is not proper. If one is quick of eye he will be deft of hand. Therefore, when we go along with the electrical testing after the mechanical installation, the first step is to carry out the sensor's installation and testing.

Note:

Sensors commonly used in the production lines include proximity switch, displacement measuring sensor, pressure measuring sensor, flow measuring sensor, the temperature and humidity testing sensor, compositions detecting sensor, image detecting sensor and many other types. We do not introduce all here. Each kind of sensor varies in the use of position, requirement, detecting distance, way of installation, electrical characteristics of the output interface. So consideration should be taken on actuating mechanism and controllers during installation and testing. As it goes, "The eyes are the window of the soul". No sharp sense, no quick action. That is to say, the automatic technology development is impossible without the sensor technology.

Engineering Competence Training

Please refer to the product manual of the YL-335B Automatic Production Line and try to explain the characteristics of each sensor. Can you understand why these sensors are used in this Automatic Production Line? What would you like to choose? What should be noted in the installation?

It seems we learned it before. Now it becomes clearer after the summary!

Task Two Control of Asynchronous Motor in the APL

It is very important! I want to learn it!
Now, I tell you how to control asynchronous motor.

Task objectives

1. Have a good command of the way of the asynchronous motor control;

2. To control asynchronous motor with the frequency converter;

3. Being capable of setting the parameter of frequency converter.

In the Automatic Production Line, there are a lot of mechanical movement controls, just like the human hands and feet, to complete the mechanical movement and action. In fact, in the Automatic Production Line, driving devices used as power sources include motors, pneumatic devices and hydraulic devices. In YL-335B, the conveyor belt's movement control in the sorting unit is performed by the AC motor. If the asynchronous machine is like a weapon, then the controller is like the maneuver. The conveyor belt in the YL-335B sorting station is powered by the three-phase AC asynchronous motor. It does not only require the speed change, but also require the direction change during the operation. The AC asynchronous motor converts the electric energy into the electromagnetic force by making use of the electromagnetic coil and then relying on the electromagnetic force to do work. So it coverts the electric energy into the rotor's mechanical movement. The AC motor's simple structure enables itself with high power and it can be used in the any places with the AC power.

Subtask One Use of the AC Asynchronous Motor

The three-phase AC motor with reduction gearbox is used in the conveyor belt in YL-335B sorting unit, as shown in Fig.2-24, so that the the conveyor belt runs at the moderate speed.

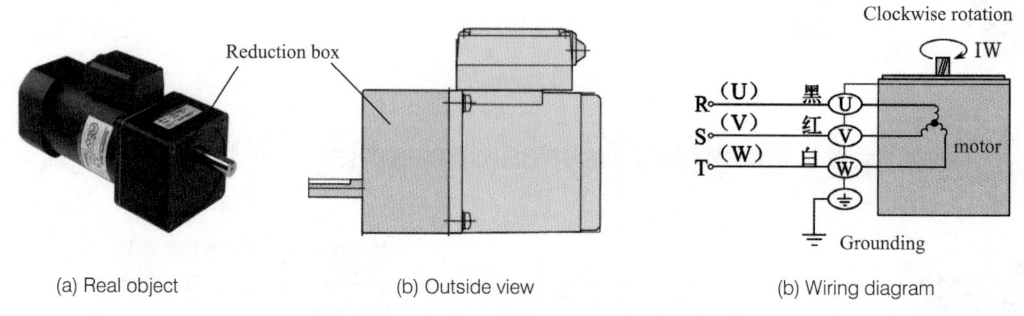

(a) Real object　　(b) Outside view　　(b) Wiring diagram

Fig.2-24 Three-phase AC reducing motor

If the frequency of the winding current of the AC asynchronous motor is f, the motor's magnetic pole number is p, then the synchronous speed (r / min) can be expressed as $n_0=120f/p$. The rotor speed of asynchronous motor n can be expressed as equation 2-1.

$$n=\frac{60f}{p}(1-s) \tag{2-1}$$

s—slip ratio.

It can be seen from the equation above that if we want to change the motor's revolutions: ① change the magnetic pole number p; ② change the slip ratio s; ③ change the frequency f.

In the control of conveyor belt in sorting unit of YL-335B, the speed control of the AC motor is completed by means of the frequency control. How could we control the conveyor belt's direction? Conventionally, we can change the phase sequence of the AC motor's power to change the direction of motor's rotation.The motor's speed and direction control in the sorting unit are completed by the frequency converter.

It should be noted that during operation if one phase of the three-phase asynchronous motor is powered off, then it becomes single-phase operation. At this time the motor will still run in the original direction. However, if the load does not change and the three-phase power becomes the single-phase power, then the current will become bigger, causing the motor overheat. Special attention should be paid to this. If one phase of the three-phase asynchronous motor is powered off before starting, it can not be started and we can just hear the humming sound. If it can not be started after a long time, it will be overheated and trouble shooting is required immediately.Please note that the shell's grounding wire must be connected to the ground to prevent personal injury caused by electric leakage.

Subtask Two Use of Universal Frequency Converter Driver

The speed and direction control of the three-phase AC geared motor in the sorting unit in YL-335B adopts Siemens' universal frequency converter MM420. Its electrical wiring is shown as Fig.2-25. The three-phase AC power is output to the AC motor through the fuse, the AC contactor, the filter (optional) and frequency converter.

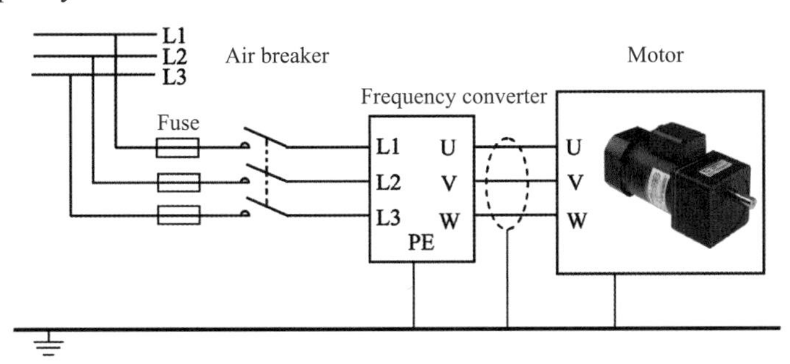

Fig.2-25 Installation wiring of the frequency converter and the motor

In the Fig.2-25, there are two points to note: the first is shielding, the second is grounding. The wiring both from the filter to the frequency converter and from the frequency converter to the motor must be the shielded ones. The shielded layer must be grounded; Furthermore, the casing of the charging equipment must be grounded as well.

In real practice, complex setting on the frequency converter should be done according to the control requirement of the project in the automatic Production Line. For more information, please refer to the relevant technical manuals. Other servo driving devices can also be chosen according to the motor's driving requirements in the Automatic Production Line, such as the transistor DC PWM drive used in the DC motor and the vector control AC frequency converter drives, etc.

Summary of Knowledge and Skill

Speed adjusting by frequency converting is the future development of the AC speed regulation, which has been widely used. The sine wave's PWM is to control the inverse transformer switch elements' of on and off according to certain rules, resulting in a group of rectangular pulse with the same amplitude but different width. Its fundamental wave is just like the sine wave voltage. Nowadays the frequency converter becomes more and more intelligent. In practice much attention should be given to the parameter setting, the connection with the external devices and the control.

Engineering Competence Training

Refer to the information about the the motor and the frequency converter of the manufacturers, sort out the precautions in the installation and testing process, and choose the appropriate diving device based on the motor.

Yeah, I have got this weapon and this maneuver.

Task Three Application of Servo Motor and Driver in the APL

Here is one kind of servo motor and driver.

Task Objectives

1. Grasp the characteristics and ways of servomotor control, the principles and electrical wiring of servo driver;

2. Being able to control a servomotor with a servo-driver;

3. Being able to set the parameters of a servo-driver.

Servo motor, also known as operating motor, is used as the actuator in the automatic control

system, to transform the received electrical signal into the angular displacement or angular velocity in a motor shaft and then to send this out. Servo motor mainly includes two kinds, DC and AC servo motor. Their main features are that, when the signal voltage is zero, there comes no rotation, and the revolution speed will fall evenly along with the torque increase. AC servo motor is a brushless motor, which can be divided into synchronous motor and asynchronous motor. The former is generally used in motion control system. Its large power range and inertia enables great power and makes it suitable for low speed running smoothly.

Since the 1980s, along with the development of integrated circuits, power electronics technology and AC variable speed drive technology, the permanent magnet AC servo drive technology booms. This AC servo system has become a major direction in the development of the contemporary high-performance servo systems.

Currently, high-performance servo systems mainly adopt permanent-magnet synchronous AC servo motor, and the control drivers mostly employ the all-digital position servo system which is faster and of accurate location. Typical manufacturers are Siemens of Germany, Cole Morgan of the United States, Yaskawa of Japan and so on. The Panasonic MINAS-A4 series servo motors and drives are being used in YL-335B.

Subtask One Getting to Know AC Servo Motor and Driver

In the delivery (transfer) unit of YL-335B, the Panasonic MHMD022P1U permanent-magnet synchronous AC servo motor is being used. The MADDT1207003 all-digital AC permanent-magnet synchronous servo drive works as a movement control device of the transport manipulator, shown as Fig.2-26.

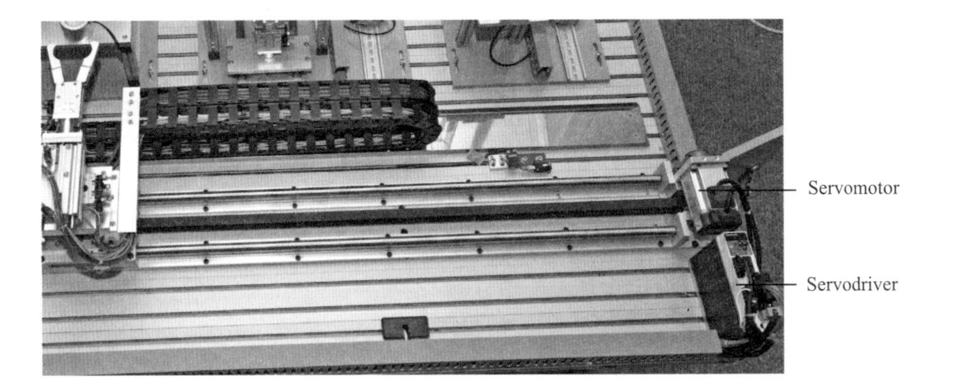

Fig.2-26 The servo motor and driver in the delivery unit in YL335B

The working principle of an AC servo motor is as the following: the rotor inside a servo motor is a permanent magnet. The U / V / W three-phase controlled by the driver forms the electromagnetic field. The rotor rotates under the effect of this magnetic field. While the motor's build-in encoder sends feedback signal to driver which compares the feedback value with target value, then adjust the rotation angle of the rotator. The accuracy of a servo motor depends on the accuracy of its encoder (number of lines). Its structure is shown as Fig.2-27. Note that the motor encoder is most likely to be damaged in a servo motor, since there are precision glass discs and optical element. Thus a motor should be protected from strong vibration and hit on the end and the encoder part.

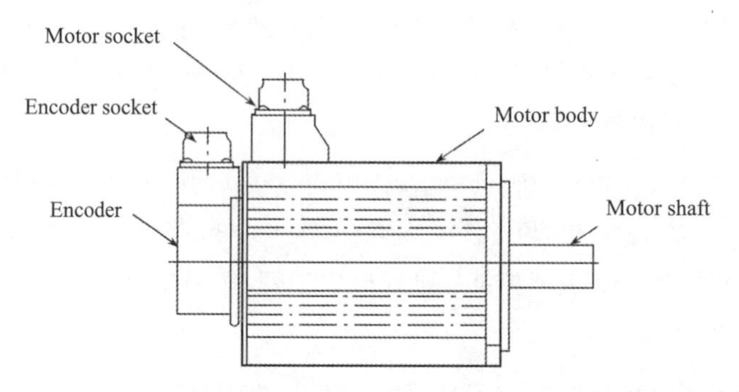

Fig.2-27 Structure diagram servo motor

The meaning of MHMD022P1U: MHMD represents that the motor type is high inertia; 02 represents that the rated power is 200 W; 2 means that the voltage specifications is 200 V; P means that the encoder is an incremental encoder, with the pulse number of 2,500 p/r, resolution rate of 10,000, output signal lines of 5 .

AC permanent-magnet synchronous servo driver is composed of servo control unit, power drive unit, communications interface unit, servomotor and the corresponding feedback detection components. Its control system structure is shown as Fig.2-28. The servo control unit includes the position controller, speed controller, torque and current controller, etc.

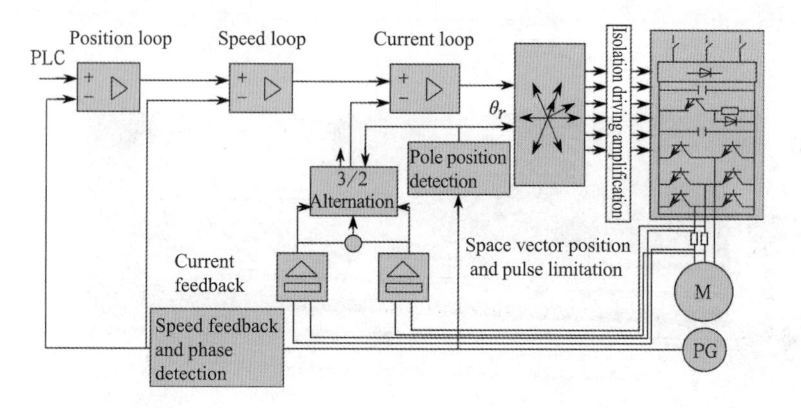

Fig.2-28 Structure Diagram of the Servo Driver

The meaning of MADDT1207003: MADDT means Panasonic A4 Series-A driver; T1 represents the maximum instantaneous output current is 10 A; 2 shows that power voltage specification is single-phase 200 V; 07 means the rated current of current monitor is 7.5 A; 003 means special pulse control. The panel graph is shown as Fig.2-29.

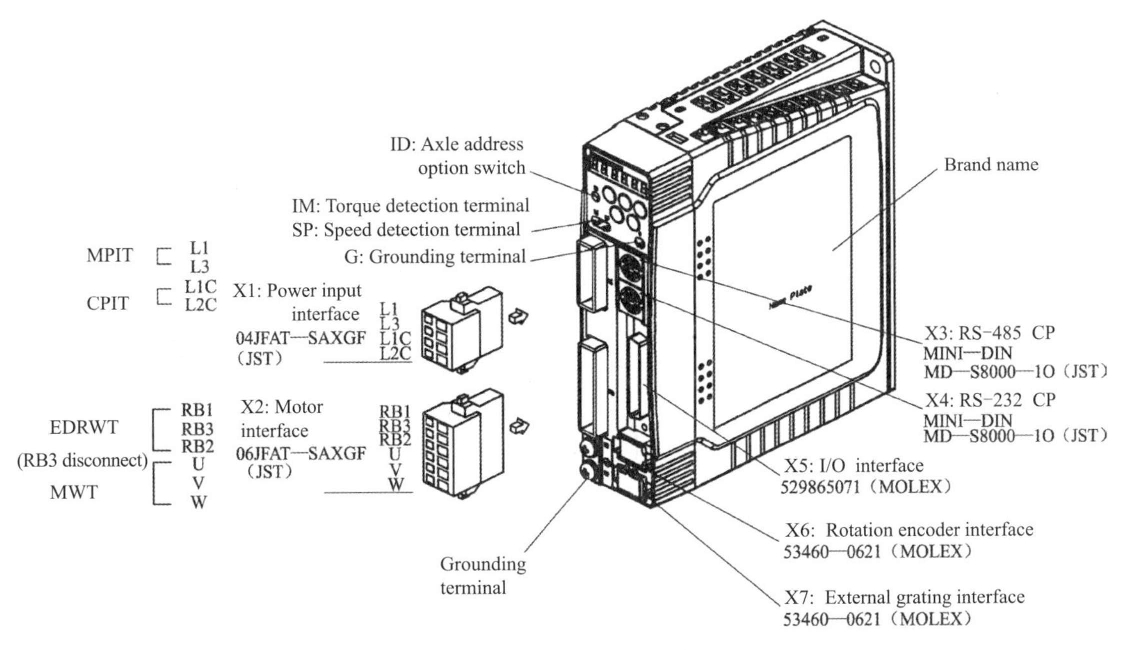

Fig.2-29 Servo driver panel graph

Panasonic servo drivers is available in seven control and running modes, that is, position control, speed control, torque control, position/speed control, position/torque, speed/torque, and closed-loop control. Position mode is to input the pulse series to run the motor positioning. Motor revolution is related to pulse series. The motor rotation angle is associated with the pulse number. Speed control mode has two ways, one way is through the input of DC −10 ～ +10 V command voltage to regulate speed, the other way is to select the internal speed set in the driver to regulate speed. Torque mode means through the input of DC −10 ～ +10 V command voltage to regulate the motor's output torque. Under this mode a speed limit is needed. There are two ways to realize this limit, one is to set the driver internal parameters and the other is to input the analog voltage.

Subtask Two Hardware Wiring of Servo Motor and Driver

The wiring between servo motors and drivers and peripheral devices are shown as Fig.2-30. Input power flows through the circuit breaker, the filter, and then directly into the control power supply input (X1) L1C, L2C. The power following the filter flows through the contactor and the reactor into the servo driver's main power supply input (X1) L1, L3. Servo driver's output power supply(X2) U, V, W connects with servo motor. The output signal from servo motor's encoder should also be connected with the driver's encoder input port (X6). The associated I/O control signal (X5) might also be connected with the PLC and other controllers as well. Servo drivers can also be connected with a computer or hand controller for parameter setting.Then ue will introduce servo drive device wiring from three aspects.

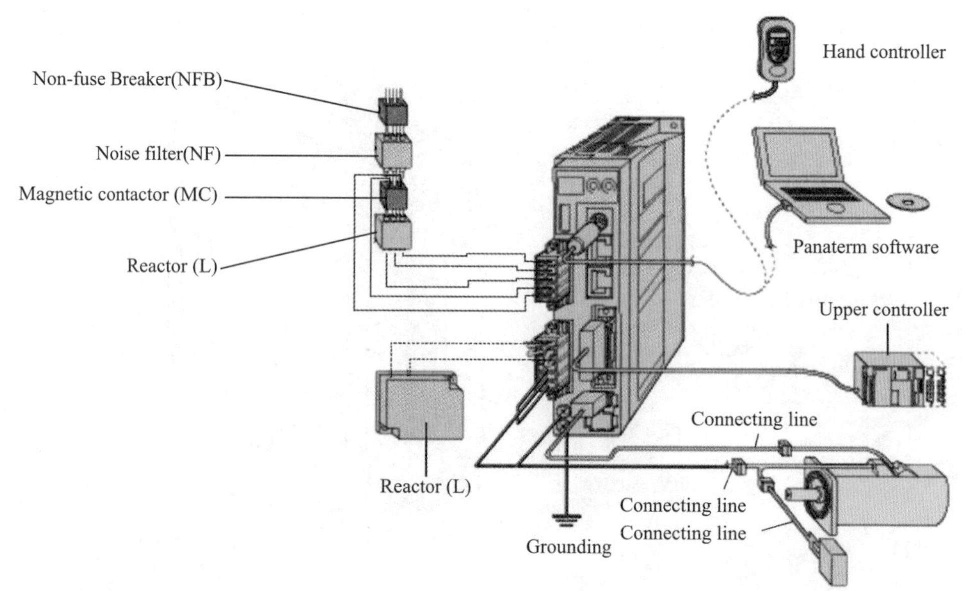

Fig.2-30 Wiring diagram between servo motor and driver and the external devices

Servo driver parameter's setting is shown as Table 2-2.

Table 2-2 Servo driver parameter's setting

Serial number	Parameter		Setting value	Functions and definition
	Number of parameter	Parameter name		
1	Pr01	LED original state	1	Shows the motor revolution
2	Pr02	Control mode	0	Position control（related code P）
3	Pr04	Invalid setting input is prohibited in travel limit	2	When left or right limit action occurs, Err38 travel limit will give prohibit ion of input error alarm. This parameter can only be modified and written successfully after power-down and restart of control power.
4	Pr20	Radio of inertias	1678	The value is automatically adjusted, see AC for details.
5	Pr21	Real time automatic gain setting	1	It automatically adjusts to conventional mode in real time and the change of load inertia is very small during operation.
6	Pr22	Mechanical rigidity selection of real-time automatic gain	1	The bigger this parameter being set, the faster is the response, but too high parameter would cause unstable.
7	Pr41	Command pulse rotational direction setting	1	Command pulse + command direction. This parameter can only be modified and written successfully after the power is off and restart of control power.
8	Pr42	Command pulse input mode	3	Command pulse + Command direction; PULS; SIGN; L Low level; H High level
9	Pr48	No.1 numerator of command pulse sub-octave	10000	Command pulse number per revolution=Encoder resolution× $\frac{Pr4B}{Pr4B \times 2^{Pr4A}}$ Encoder resolution is 10,000(2,500p/r×4), then required command pulse number per revolution is Encoder resolution is 10,000 (2,500p/r×4), then required command pulse number per revolution is=10,000× $\frac{Pr4B}{Pr4B \times 2^{Pr4A}}$ =10,000× $\frac{5,000}{10,000 \times 2^0}$ =5,000
10	Pr49	No.2 numerator of command pulse sub-octave	0	
11	Pr4A	Numerator radio of command pulse sub-octave	0	
12	Pr4B	Denominator of command pulse sub-octave	5000	

For descriptions and settings of other parameters, please refer to the manuals of servo motors and driver of APL in the disk provided.

 Summary of of Knowledge and Skills

In YL-335B, the AC servo motor is the movement actuator of the transport unit. Its function is to convert electrical signals into linear displacement or speed of the manipulator. Servo motors are divided into two major categories, the AC servo and the DC servo motors. The AC servo system has become a major development trend of high-performance servo systems now. The generally used AC servo motor is composed of permanent magnet synchronous motor and coaxial optical encoder. The accuracy of the built-in encoder determines the control precision.

Note: An AC servo driver is composed of the servo control unit, the power drive unit, the communications interface unit, the servo motor and the corresponding feedback detecting devices. The servo control unit includes the position controller, speed controller, torque controller and so on. Students are supposed to master the electrical characteristics of servo motor, servo driver, to fully understand the external ports' function of the servo driver and be able to connect the wires correctly and set the control parameters of the servo driver accurately in this task.

 Engineering Competence Training

Servo drivers involve many parameters and complicated external ports. Consult the manufacturer information about AC servo motor and driver, based on the requirements of sample tests in the disc to learn the functions of all the external ports. Sort out the related parameters' function of a servo driver and try to test the servo motor and servo driver manually.

 Task Four Application of Pneumatic Technology in the APL

Task Objectives

1. Master the functions and characteristics of the commonly used pneumatic components;

2. Being able to use the pneumatic components to form the pneumatic system and connect the gas path.

In YL-335, there are a number of pneumatic components, including air pump, filter pressure-reducing valve, one-way electrical control valve, two-way electrical control valve, cylinder, bus-bar and so on. Among these, the cylinder adopts pen-type cylinders, thin cylinders, rotary cylinders, double-rod cylinders and finger-type cylinders, totally 17 of five types. As shown in Fig.2-31.

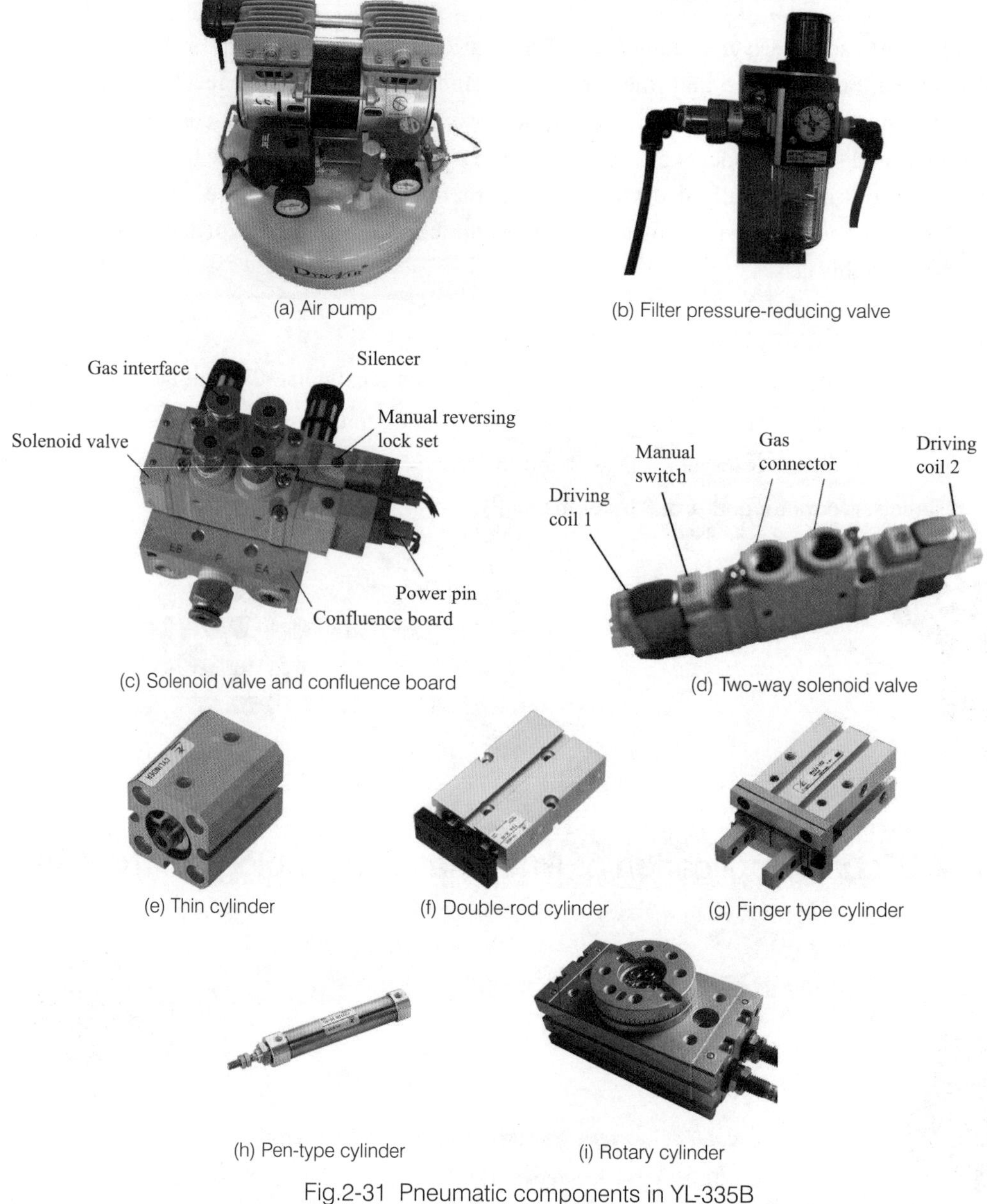

(a) Air pump (b) Filter pressure-reducing valve

(c) Solenoid valve and confluence board (d) Two-way solenoid valve

(e) Thin cylinder (f) Double-rod cylinder (g) Finger type cylinder

(h) Pen-type cylinder (i) Rotary cylinder

Fig.2-31 Pneumatic components in YL-335B

The above figure contains the following four parts: gas supply devices, control components, actuators and auxiliary components.

1. Gas supply device: It transforms the mechanical energy from the prime motor into the air pressure energy. Its main device is the air compressor, Fig.2-31 (a) shows the air pump.

2. Control components: Control components are used to control the air pressure, flow rate and flow direction, so as to ensure that the actuator has a certain output power and speed in accordance with the designed procedure. Fig.2-31 (c), (d) shows the solenoid valve.

3. Actuators: Actuators are energy-conversion devices to transform air pressure energy into mechanical energy. Fig.2-31 (e) (f) shows the various cylinders' types.

4. Auxiliary components: Auxiliary components are devices used to ensure the normal operation of the air system. Such as filters, dryers, air filters, silencers and oil atomizer, etc. (As shown in Fig.2-31 (b)).

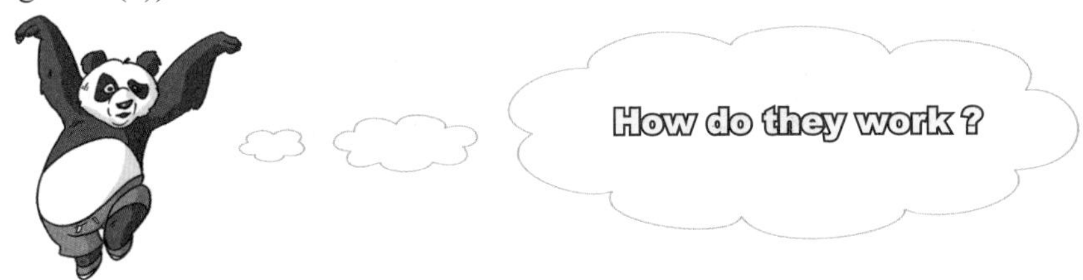

Note: Pneumatic system is used for energy and signal transfer by taking the compressed air as the working medium. And it transfers the mechanical energy from motor or other prime motor into the air pressure energy. Then it transforms pressure energy into mechanical energy by way of the actuator unit with the support of the control components and the auxiliary components. Thus completing the linear or rotary movement, meanwhile do work towards outside.

Subtask One Getting to Know Pneumatic Pump

Fig.2-32 shows the air pump which generate aerodynamic force, including air compressor, pressure switch, overload safety protector, gas tank, pressure gauge, air supply switch, and main pipe filter.

Fig.2-32 Introduction of air pump components

The gas supply device mentioned above is a set of device which is used to generate compressed air with sufficient pressure and flow rate and then to do purification, handling and storage. It mainly consists of the following components: air compressor, aftercooler, oil remover, gas tank, dryer, filter, and gas pipeline.

Subtask Two Getting to Know Pneumatic Actuating Components

The commonly used actuator components in pneumatic system are cylinders and gas motors. Cylinder is for the linear reciprocating movement; gas motor is used for continuous rotary movement. In YL-335B, only cylinders are used, including the pen type cylinder, thin cylinder, rotary cylinder, double-rod cylinder, and finger cylinder, as shown in Fig.2-33 below.

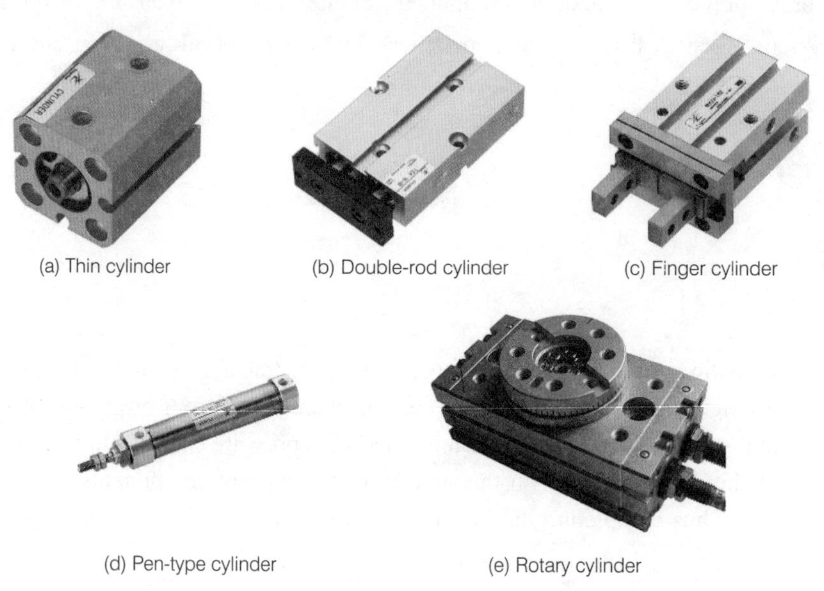

(a) Thin cylinder (b) Double-rod cylinder (c) Finger cylinder

(d) Pen-type cylinder (e) Rotary cylinder

Fig.2-33 Cylinders used in YL-335B

A cylinder mainly consists of cylinder body, piston rod, front and back lids, seals and so on, Fig.2-34, is a structure of a common single-piston, dual-function cylinder.

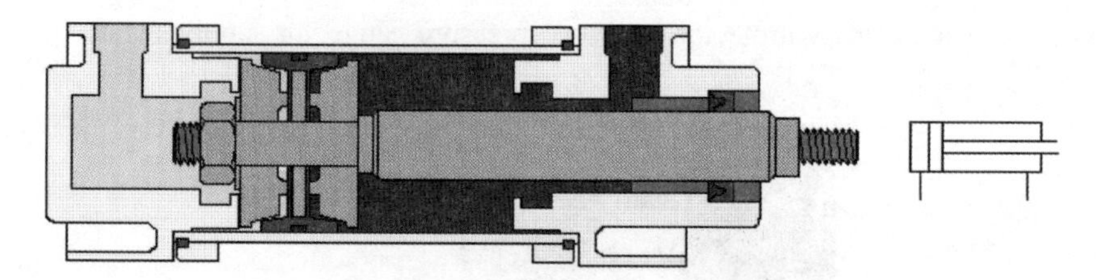

Fig.2-34 Structure of a common single-piston and dual-function cylinder

The so-called dual-function means the piston's reciprocating movement is pushed by compressed air. In a single-out piston power cylinder, due to larger space on the right of the piston, when the air pressure works in the right, it will provide a slow and a big work force stroke. When it returns, due to the smaller space on the left of piston, the speed is faster and the force becomes less. This kind of

cylinder is most widely used, usually in packaging machinery, food machinery, processing machinery, etc.

The main components of material turret is the pneumatic swing platform. The rotary movement is realized by way of the linear cylinder driving rack and pinion. The rotation angle can be adjusted between 0°～90° and 0°～180°. And the magnetic switch can be installed to detect the rotating position signal. for the structure which needs to alter the direction and position. As shown in Fig.2-35.

Fig.2-35 Pneumatic swaying platform

The pneumatic swing platform used in YL-335B has a swing rotary angle which can be adjusted in the range of 0°～180°. When you need to adjust the rotary angle or adjust the swing position accuracy, you should first release buckle screw on the adjusting bolt, through rotating the adjustment screw of bolt to change the rotation angle on the rotary convex platform. The adjusting bolt 1 and bolt 2 are used to control left and right angle adjustment accordingly. After adjusting the swing angle, the buckle screw and matrix should be fastened to prevent the loose of screws. The loose of screws will cause the reduction of the precision.

There are many different types of cylinders, so as to the way of classification. Usually the cylinders are classified according to direction of compressed air acting on the piston surface, structure and way of installation. It can also be classified by the size of cylinder. Usually cylinder with 2.5 ~ 6 mm bore is known as mini cylinder, 8~25 mm is little cylinder, 32~320 mm as the medium cylinder, more than 320 mm as the large cylinder. Regarding the way of installation, there are fixed cylinder and swing cylinder. According to the way of lubrication, cylinders can be divided into oil feeding cylinders and non-oil feeding cylinders. Regarding the driving mode, cylinders can be divided into single-acting cylinders and double-acting cylinders.

Subtask Three Getting to Know Pneumatic Control Components

The pneumatic control components used in YL-335B involves pressure control valves, direction control valves, flow rate control valves according to the use and function.

(1) Pressure control valve

The pressure control valves used in YL-335B mainly include reducing valves and overflow valves.

① Pressure-reducing valve is used to reduce the pressure from the air compressor to meet the need of each pneumatic device and to keep the pressure stable. Fig.2-36 is a direct pressure-reducing valve.

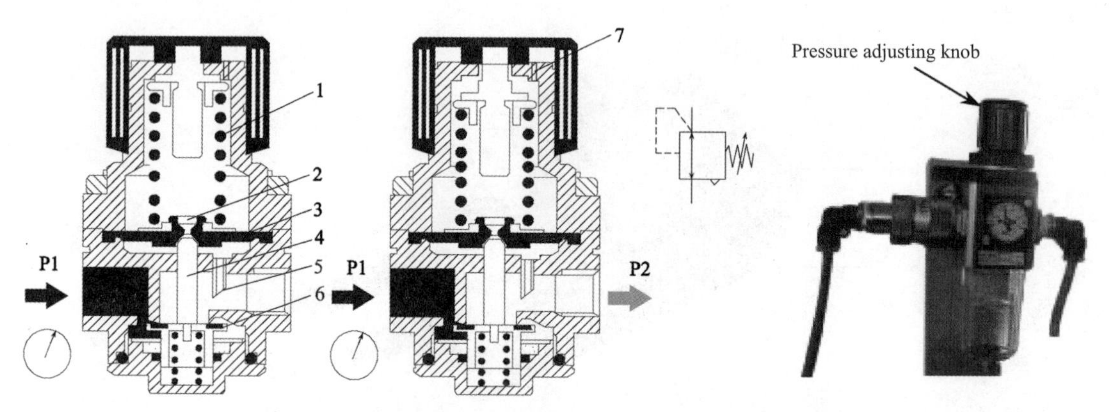

Fig.2-36 Structure and physical picture of reducing valve

1-Pressure adjusting spring; 2- Overflow valve; 3-Diaphragm; 4-Valve rod; 5-Feedback guide rod; 6-Main valve; 7-Overflow outlet

② The function of the overflow valve is to exhaust gas automatically in order to reduce the system pressure so as to ensure the system safety when the system pressure exceeds the set value. For this reason it is also called the safety valve. Fig.2-37 shows the principle diagram of the safety valve.

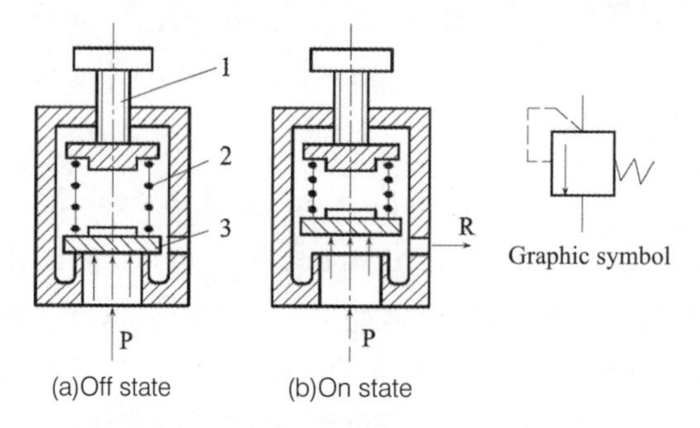

Fig.2-37 Functional diagram of safety valve

1-Knob; 2-Spring; 3-Piston

(2) Flow control valve

The flow control valve in YL-335B mainly refers to the throttle valve.

Throttle valve reduces the air flow cross-section so as to increase the air flow resistance, and accordingly the air pressure and flow would decrease. As shown in Fig.2-38, there is an adjustment

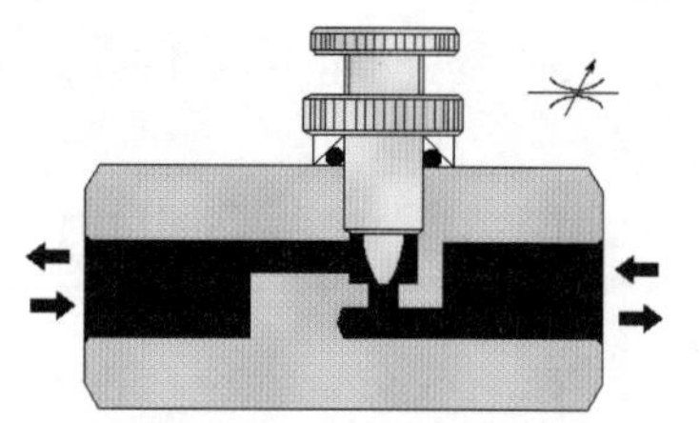

Fig.2-38 Structure diagram of the throttle valve

screw on the valve body, you can adjust the opening of the throttle valve (stepless regulation), and maintain the same opening. This kind of valve is called adjustable open throttle valve.

Adjustable throttle valve is often used to adjust the motion speed of the cylinder piston. It can be installed directly on the cylinder. The throttle valve has two-way throttling function. When you use the throttle valve, the throttling area cannot be too small, since the condensed water and dust of the air would block the pass and cause the change of throttling.

To ensure the cylinder motion smoothly and reliably, limited stretch throttle valve is installed on action air port. The throttle valve is supposed to regulate the cylinder motion speed. There is a quick adapter of the air pipe on the throttle valve , as long as we put the air pipe with suitable diameter into the quick adapter it would fit at once. It is very convenient. Fig.2-39 shows the front view of the cylinder which is installed limited stretch throttle valve with quick adapter.

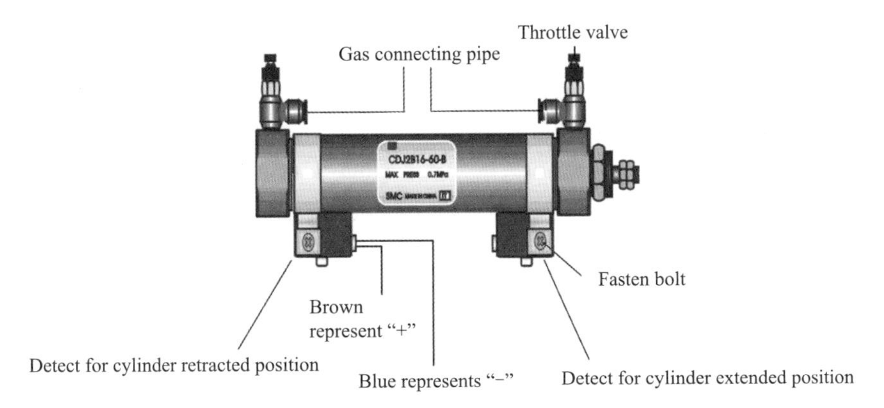

Fig.2-39 Cylinder with throttle valve

Fig.2-40 (a) is a schematic diagram of connection and adjustment of a double action cylinder with two limited stretch throttle. When you adjust the throttle valve B, the extending speed of the cylinder will be adjusted. When you adjust the throttle valve A, the retracting speed of the cylinder will be adjusted.

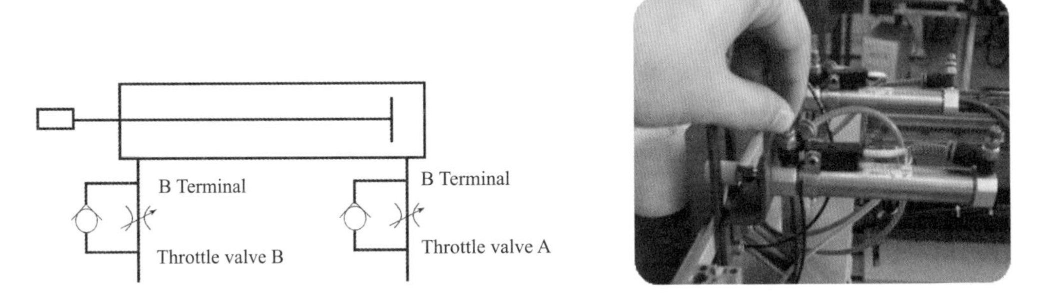

Fig.2-40 Connecting and adjusting of throttle valve

(3) Direction control valve

A direction control valve is used to change the air flow direction or to determine the on-off state. Usually use solenoid valve.

Solenoid valve is used to change the air flow direction by way of when the electromagnetic coil being charged with electricity, static magnetic iron producing electromagnetic suction to the dynamic iron core, resulting in the valve core switching. Fig.2-41 shows the working principle of a single electrical control, two-way, three-port reversing solenoid valve.

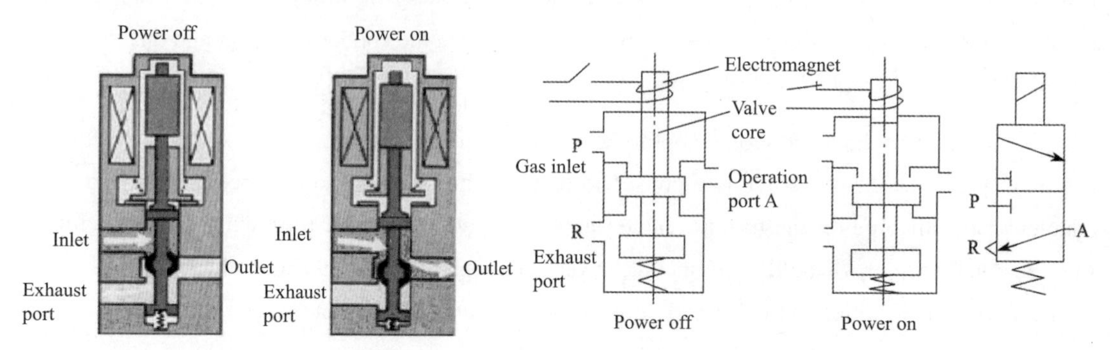

Fig.2-41 Working principle of single electrical control reversing solenoid valve

The so-called "position" refers to change the air direction and valve core different operation positions with relative way to the valve body. "way" refers to the passes between the direction control valve and the system. The number of the port determines the "how many ports". In Fig.2-41, there are only two operation positions and also there is gas inlet P, the operation port A and exhaust port R, so it is called two-position three-way valve.

Fig.2-42 lists the graphic symbols of two-position three-way, two-position four-way and two-position five-way single electrical control reversing solenoid valve.The number of square in the diagram stands for the number of position, and the "┬" and "┴" symbols in the square indicate that the interfaces are mutually independent.

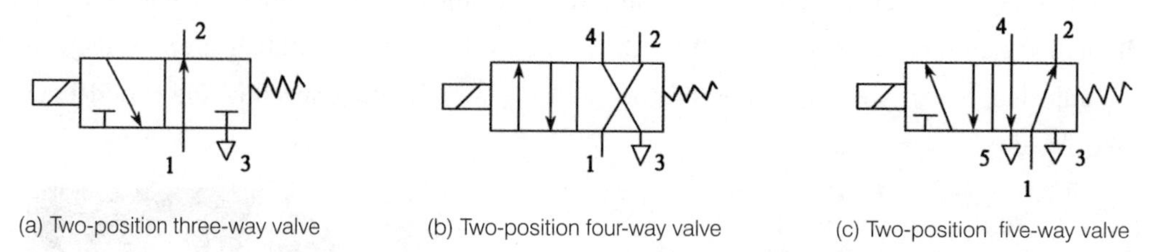

(a) Two-position three-way valve (b) Two-position four-way valve (c) Two-position five-way valve

Fig.2-42 Graphic symbols of some single control reversing solenoid valves

All the actuating cylinders in working unit in YL-335B are double action cylinders, thus the control solenoid valves are supposed to have two working ports, two exhausting ports and one air supply port. So the solenoid valves are all two-position five-way solenoid valves.

In YL-335B, all the valves are connected in the form of valve set, which integrates a number of valves and silencers, and bus-bar forming a set of control valve, meanwhile each valve functions is independent.

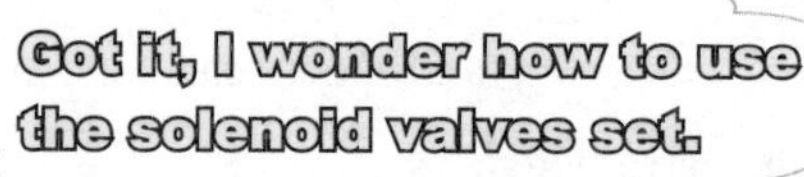

Taking the feeding unit for example, it adopts two two-position five-way single control solenoid valves. The two solenoid valves are with manual reversing switch and locking button. There are two positions of lock (LOCK) and open (PUSH). When you get the lock knob to LOCK position with a small screwdriver, the manual control switch will get concave and the manual operation no longer works. Only on PUSH position, it is available for tools to push down. When signal is "1", it means that the electromagnetic signal of this side is "1". In normal state, the manual switch signal is "0". During the testing, it is available to use manual control switch to control the valve and to realize the control of corresponding gas path. Thus it accordingly changes the control of pushing cylinder and some other executing units to realize the objectives of the testing.

Two solenoid valves are installed on the bus-bar. The two ends of the exhaust port on the bus-bar connect with the silencers. The function of silencers is to reduce the noise when the compressed air is emitted into the atmosphere. The integration of several valves, silencers and the bus-bar is called valve set. Each valve on this valve set works independently. The structure of the integrated valve set is shown as Fig.2-43.

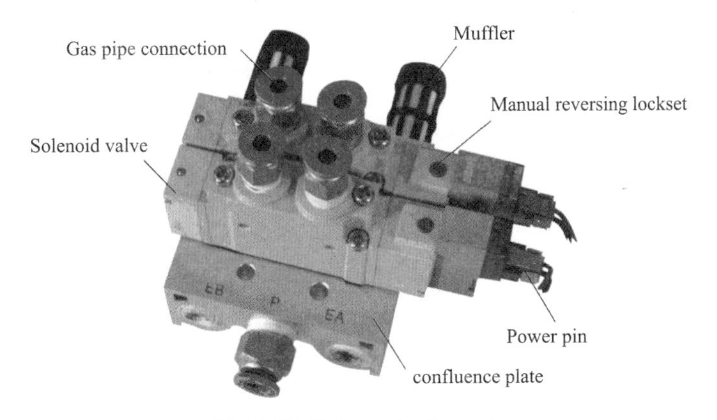

Fig.2-43 Solenoid valve set

In the transport station, the double action cylinder with pneumatic gripper is controlled by a two-position five-way double action solenoid valve. It is capable of maintaining function for each workstation's grabbing and carrying. The working principle of double action solenoid valve is similar to double steady-state trigger which means the output state is determined by the input state. If the output state identified, even without the input state, double action electrical control solenoid valve will maintain the state as before. Double action electrical control solenoid valve is shown as Fig.2-44.

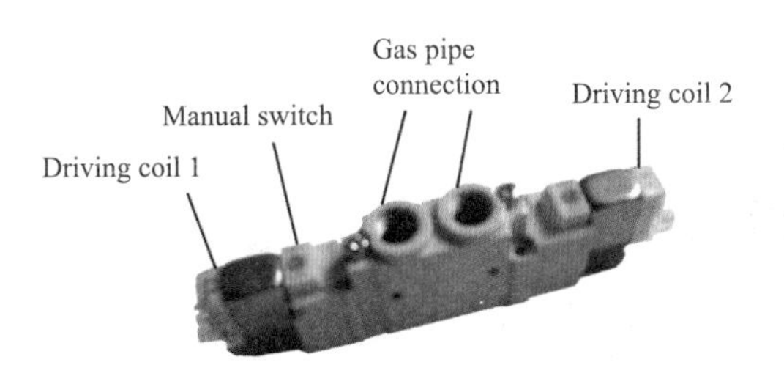

Fig.2-44 Diagram of double action electrical control solenoid valve

Double action electrical control solenoid valve is different from single action electrical control solenoid valve. Regarding the single action electrical control solenoid valve, it can be reset with the valve core being affected by the spring force even though without control signals. Regarding the double action electrical control solenoid valve, the valve core position is decided by previous electrical control signal while no electrical control signal on both ends.

Double-rod cylinder is a double action cylinder controlled by a two-position five-way single electrical control solenoid valve, which used to control the gripper's extending and retracting movement.

Rotary cylinder is a double action cylinder controlled by a two-position five-way single electrical control solenoid valve. It is used for controlling the arm rotation angle in the range of positive and negative 90° . The cylinder rotation angle can be adjusted between 0°~180°. The adjustment is realized through two fixed buffers below the throttle.

Note: Two electrical control signals of double electrical control solenoid valves cannot be "1" at the same time, which means both coils in the control process cannot be energized at the same time. Otherwise, it may cause burning of the electromagnetic coils. Of course, in this case the valve core position is uncertain.

Lift cylinder is a double-acting cylinder controlled by a two-position five-way, one-way solenoid valve, which used for the lift of the manipulator.

The running speed of the cylinder mentioned above is regulated by the inlet throttle valve by way of regulating the inlet gas.

Now a solenoid valve is damaged, so it is necessary to replace. Go ahead and have a try, can you figure out how to install a solenoid valve?

① Cut off the gas and remove the damaged valve with a screwdriver, as shown in Fig.2-45.

② Install the new solenoid valve with a screwdriver, as shown in Fig.2-46.

Fig.2-45 Bus-bar with solenoid valve removed

Fig.2-46 Install solenoid valve

③ Insert the control connector into solenoid valve, as shown in Fig.2-47.

④ Insert the gas circuit pipe into the quick connector of thc solenoid valve, as shown in Fig.2-48.

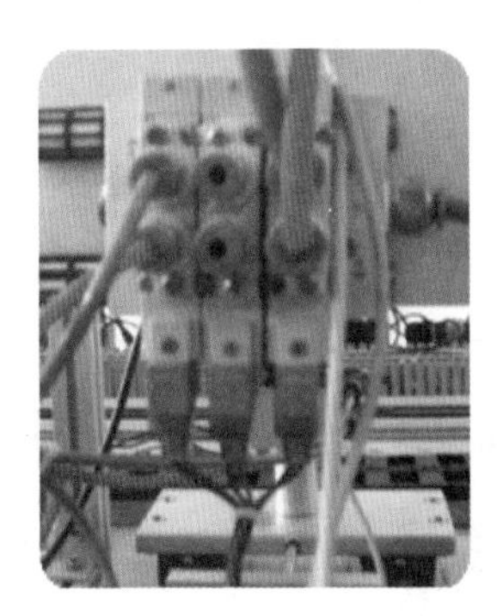

Fig.2-47 Connect solenoid valve circuit

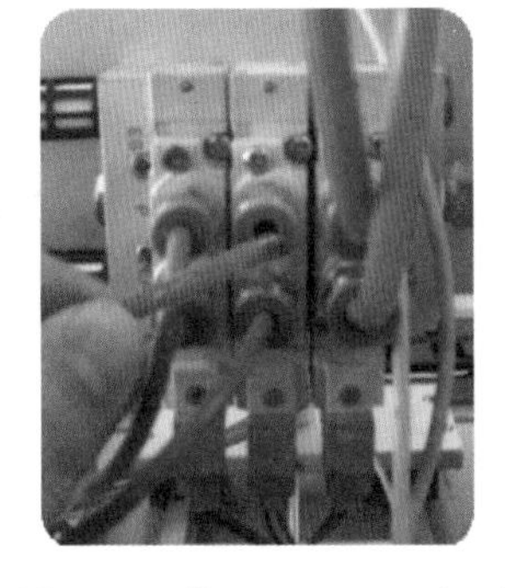

Fig.2-48 Connect gas circuit

⑤ Turn on the gas and adjust with manual switch to check the cylinder's movement.

Summary of Knowledge and Skills

Note: Basic components in the pneumatic system: generation of compressed air, compressed air supply and the consumption of compressed air. Pneumatic technology has many advantages compared with mechanical drive, electric drive and hydraulic drive. Regarding way of driving, cylinder, as a linear driver, can choose and form its motion trajectory in any position with easy installation and maintenance. Besides, it has the following advantages: an inexhaustible supply of working medium, no air pollution, low cost, low-pressure levels, safety use, fire-proof, explosion-proof and moisture-proof.

Engineering Competence Training

Referring to the pneumatic manual and think how to select pneumatic components. Get to know the current domestic and international major manufacturer of pneumatic components, the development of pneumatic technology and the application area and industries. Try to write a review on your own.

Great accomplishment with little effort! I can manage pneumatic technology too.

Task Five Application of PLC in the APL

Task Objectives

1. Grasp PLC working principle, features of external interface, choosing principle of input and output port, common commands;

2. Being able to analyze the process requirements of the control system, determine the number of input and output of digital and analog;

3. Being able to use common commands and to make control system program.

At every station of YL-335B Automatic Production Line, there is a Siemens S7-200 series PLC (Programmable Logic Controller). It is just like our brain, thinking each movement, each manoeuvre, each way. It can also command the manipulator, action of gas gripper according to the program. It is the core components of the Automatic Production Line. So, what is exactly a PLC?

Note: PLC is a digital computing operating system, designed for applications in industrial environments. PLC is developed based on electrical control technology and computer technology. It gradually becomes a new type of industrial control device with microprocessor as core which integrates automatic, computer and communications technology.

PLC used in YL-335B see Table 2-3

Table 2-3 PLC used in YL-335B

PLC type/specification	Applied unit	Feature
S7-200-224CN AC/DC/RLY	Feeding unit	14/inputs, 10/relay output
S7-200-226 AC/DC/RLY	Assembly unit	24/inputs， 16/ relay output
S7-200-224 AC/DC/RLY	Processing unit	14/inputs,10/ relay output
S7-200-224 XP AC/DC/RLY	Sorting unit	14/inputs,10/ relay output，including 3 analog I/O points，2 input/1 output
S7-200-226 DC/DC/DC	Delivery unit	24/inputs， 16 point transistor output

S7-200 series PLC belongs to hybrid PLC, which is composed of PLC host and expansion modules. The PLC host is composed of some basic modules like CPU, memory, communications circuit, basic input and output circuit and power. It is an integral PLC which can realize the control function individually. It includes the minimum composing unit needed by a control system. Fig.2-49 is the shape structure of S7-200CPU module, which combines a microprocessor, an integrated power and digital I/O Input/Output in a compact enclosure.

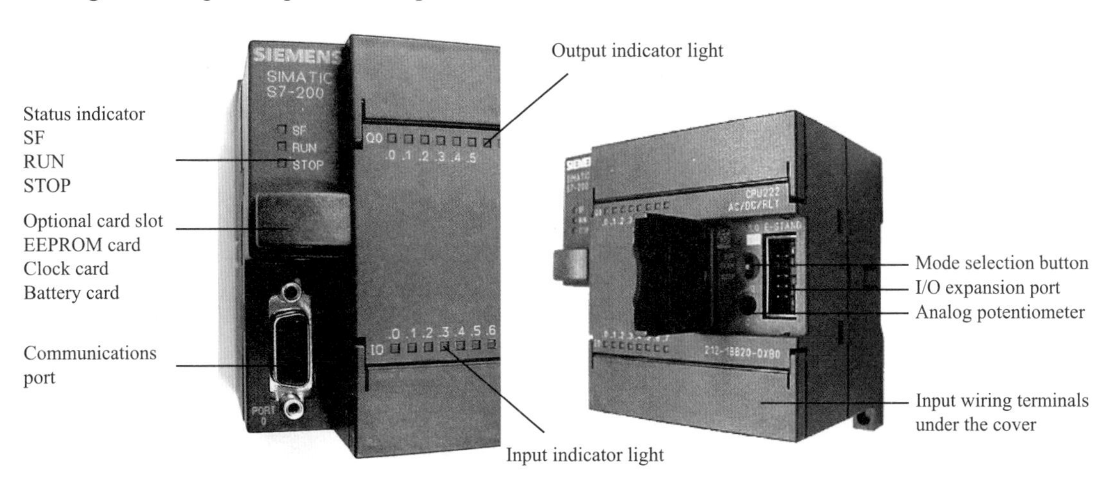

Fig.2-49 Shape structure of S7-200CPU module

Fig.2-50 shows the outer view of the PLC. Though it looks quite differently with a common computer, the PLC is just like a computer which intensifies I/O function and can conveniently connect with the control object in inner structure. The system structure of the PLC basically consists of hardware and software.

The hardware of PLC is composed of CPU, memory, input interface, output interface, communications interface, power and so on. The software of PLC consists of system program, user's program,etc.

In the internal structure, CPU module is composed of CPU, memory, input port, output port, communications interface, power and so on. Each part has different function. Like the CPU of common computer, the function of CPU of the PLC system is just like central nerve system of human beings.

1. Switching Input/Output Port

Input interface transforms digital signal or analog signal from button, stroke switcher sensor into digital signal and then delivers to CPU. Switching input often named "binary input" or "DI (Digital Input)" in engineering.

Switching Input port transforms the connected and disconnected signal from the external circuit such as button, stroke switch or sensor into digital signals1(high level), 0(low level) that can be identified by PLC and then delivers to CPU.

In Fig.2-50, The thick dashed lines represents the internal circuit. The external input forms circuit (loop) by the switch connected on the input, external power via common and PLC internal circuit. The internal circuit converts the on and off of the external switch into 0(low level)and 1(high level) signals which can be identified by CPU through photoelectric coupler. When the NPN output sensor connects with the S7-200 PLC input port, it adopts the sourcing input. When the PNP output sensor connects with the S7-200 PLC input port, it adopts the sinking input. Shown as Fig.2-51.

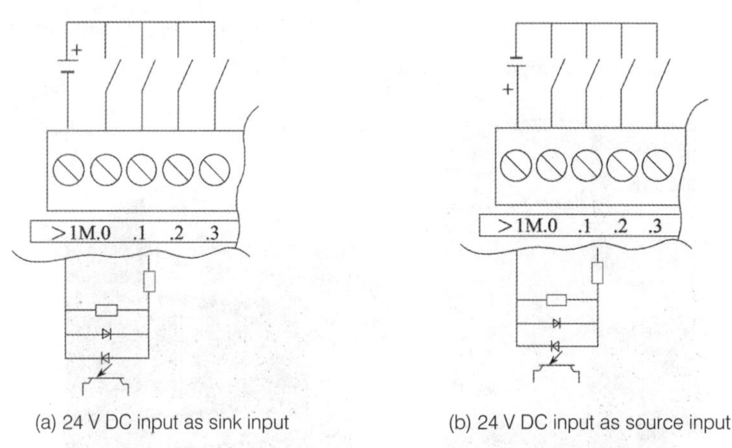

(a) 24 V DC input as sink input
(b) 24 V DC input as source input

Fig.2-50 Wiring diagram of S7-200PLC input module

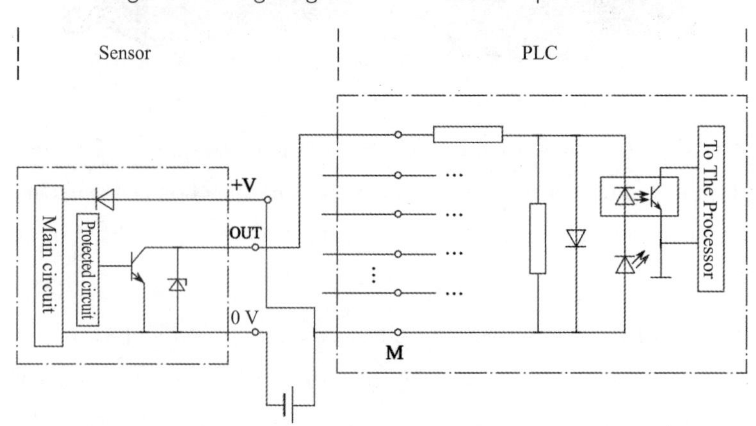

Fig.2-51 Connection between NPN output sensor and PLC

The input signal power can be supplied by the user. The DC input signal power can be supplied by PLC itself. Usually 8 or 4 input share one common terminal. The site input supplies a pair of switch signals: "0" or "1" (with or without contact is available). Every input signal will first pass through the photo isolator, filter and then be sent into the input buffer and waits to be sampled by the CPU. Each input signal will be shown by LED to indicate whether the signals reach the input terminal of PLC.

The output interface will convert the digital signals output by CPU into signals that can drive the external circuit, which can be divided into digital output and analog output. The switch output module converts the "0", "1" operation logic signal carried out by CPU into power contact output to drive the external load. Different switching output module port has different features. According to the usage

of power, it can be divided into DC output module, AC output module and AC-DC output module. According to the types of switch devices, It is divided into field effect transistor output, relay output and so on. The load type, size and corresponding time drove by them are different too. We can select needed module according to requirement. After selecting the module, we'll discuss how do you use the different modules.

Fig.2-53 shows the circuit of S7-200 series PLC output unit. 24 V DC CPU221, CPU 222, CPU224, CPU226, CPU224XP, etc 24 V DC output adopt the source output mode shown as Fig.2-52(a), CPU224XPP 24 V DC output adopt the signal flow mode shown as Fig.2-52(b), while relay output mode is shown as Fig.2-52(c).

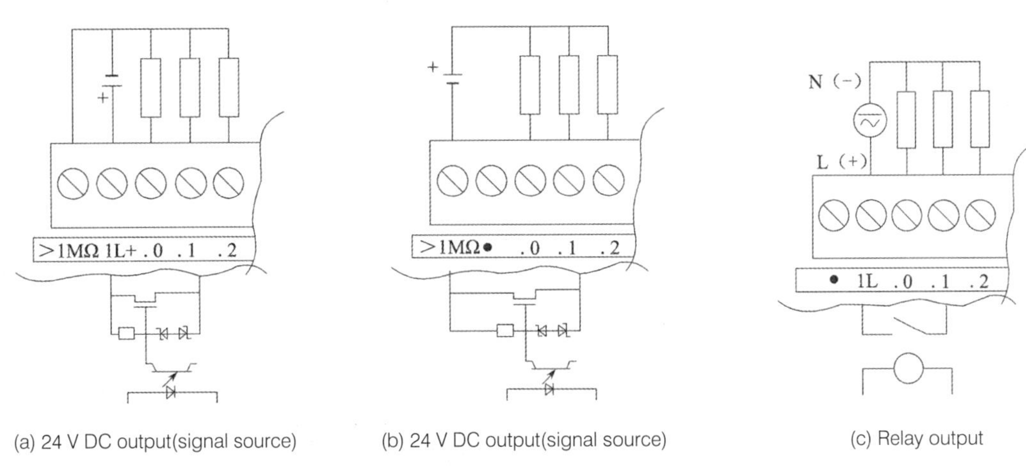

(a) 24 V DC output(signal source) (b) 24 V DC output(signal source) (c) Relay output

Fig.2-52 Wiring diagram of S7-200PLC output mode

The control of solenoid valves in feeding, processing, assembling and sorting units is necessary in YL-335B. It adopts the relay output PLC.

In the delivery unit of YL-335B, the PLC adopts the transistor output because high-speed pulse driving stepping motor and servo motor are needed to output. Based on the above, Siemens S7-226 DC/DC/DC PLC is selected.

PLC input and output interface all adopt photoelectric isolation which realizes electrical isolation of the external circuit from internal circuit to decrease electromagnetic interference.

The number of input and output interface is a key technical index for PLC. Some experts divide PLC into large, medium and small according to the I/O points.

It is important to determine the function of each I/O point in installation and testing. In the real engineering , there should be some margin allowed for the I/O points.

2. Analog I/O Module

To get the data sample of the analog, or control the position through output analog, A/D and D/A modules are essential. A/D module converts the analog quantity like voltage, current into digital quantity, while D/A is just the opposite, which converts the digital quantity into analog quantity like current, voltage signal. In the sorting unit of CPU224XP, there are two A/D, one D/A. The interface circuit is shown as the following Fig.2-66. A+、B+ are analog input quantity input interface, M is the common terminal. The input voltage range from −10 V to +10 V. The resolution rate is 11 bits plus 1 sign bit. Data format full range from -32,000 to +32,000. And the analog input image register are AIW0, AIW2. In Fig.2-53, there is a monopolar analog quantity output, which can select power output or voltage output. I is the current load output port, V is the voltage load output port. The output current range is 0～20 mA. The output voltage range from 0 V to 10 V. The resolution rate is 12 bits, data format range from 0 to 3,2767, the corresponding analog quantity output image register is AQW0.

3. Communications Interface

S7−200 PLC integrates one or two RS−485 communications interfaces that can be used as PG(programming) interface and also can be used as OP(operation terminal) interface, for example, connecting some HMI(Human-Machine interface) equipments. Or support free communications protocol and PPI communications protocol.

4. Power

S7−200 local unit has an internal power which provides a 24 V DC power for local unit, expansion module, shown as Fig.2-54. Each S7−200 CPU module provides a 5 V DC and 24 V DC power. The following two points should be noted:

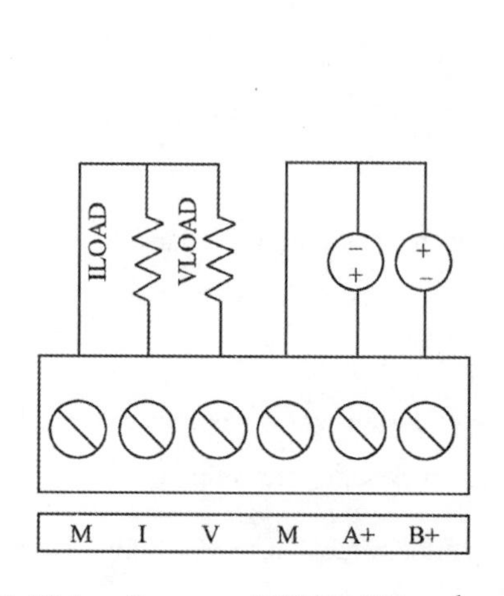

Fig.2-53 Wiring diagram of CPU224XPanalog passage

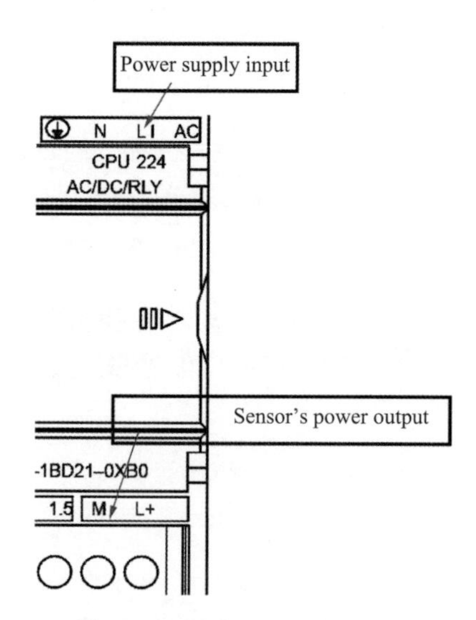

Fig.2-54 PLC power diagram

① CPU module has a 24 V DC sensor power that provides 24 V DC for local input and the expansion module relay coil. If the power exceeds the rated power 24 V DC of CPU module, you can add an external 24 V DC to provide for the expansion module.

② When any expansion module is connected, CPU module also provides a 5 V power for it. If the requirement of 5 V power of the expansion module exceeds the rated power of CPU, you have to dismount the expansion module until the requirement is within the prefixed power.

 Summary of Knowledge and Skills

 Note: The application of PLC is everywhere in our daily life. We only give two cases of YL-335B. The key points of hardware is to grasp the input and output interface properties of PLC, programming and way of adjusting. PLC instructions are very abundant. We introduce the servo stepping motor drive controlled by the output pulse with PTO pulse instruction, the usage of high-speed counter, and the realization of measurement of the displacement. We also introduce the powerful programming instruction tool guide provided by STEP7-Micro/Win. Please refer to the manuals of application method of other instructions in the disk provided.

 Engineering Competence Training

Referring to the S7-200 system manual, and think how do you write the control program of a servo(stepping) motor or electric machine, rotation speed measuring program, and arranging program testing steps and main points, writing technical files.

 Task Six Application of Communications in the APL

 Task Objectives

1. Grasp PPI communications interface protocol used in PLC and network programming instructions;

2. Grasp the installation, programming and testing of PPI communications network

In modern APL, the controlling devices in different working stations do not work independently.

For example, the five working stations work as a whole in YL-335B by way of communications means and information interexchange to form an integration and finally to improve the control and reliability of the equipment to realize “Central Handling and Separate Control”.

As one important part of automatic controlling devices, PLC can also provide us with powerful communications capacity. We can realize data exchange between PLCs through PLC communications interfaces. This task here is to learn how to use PPI communications technology in PLC.

Subtask Getting to Know PPI Communications

1. Basic Knowledge of Communications

The function of communications technology is to realize data interchange between different devices. PPI (point to point) is the serial communications from point to point. Serial communications transmit a binary number once. Thus the transmitting speed is slow. Because the wiring is short, it can transmit data in long distance. PLC-200 has its own serial communications interfaces.

2. Communications Protocol

In order to realize the communications between any devices, the both sides of communications should have an agreement on the mode and way of communications. Otherwise each side can not deliver or receive data. There are two points which need to know regarding the interface standard. First, the hardware, that is to say the number of hardware wiring, the representation of signal levels and shapes of communications connectors, etc. Second, the software, that is to say how to comprehend the definition of data delivery and receiving, and how does one side requires the other side to deliver data, which is called communications protocol.

S7-200 series PLC with communications port is the PPI communications protocol specified by Siemens. Hardware interface is RS-485 communications interface.

RS-485 only has one pair of balance differential signal used for delivering and receiving data. It is half-duplex communications mode.

Using RS-485 communications interface and connecting circuit can form a serial communications network to realize a distribution control system. The composition of a network is at most 32 substations(PLC). To improve the network anti-interference capability, two resistors are paralleled on two ends of the network.Generally the value is 120 Ω. Network wiring is shown in Fig.2-55.

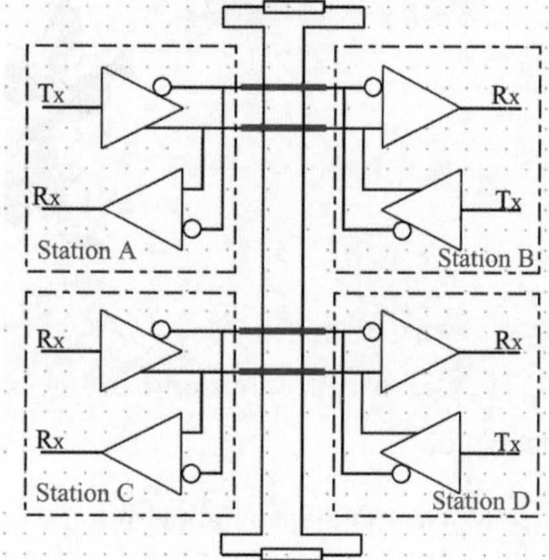

Fig.2-55 RS-485 Network wiring diagram

The communications distance of RS-485 can go as far as 1,200 m. In the RS-485 communications network, in order to distinguish device, each device has

a number, which is known as an address. The address must be unique, otherwise it would cause communications chaos.

3. Communications Parameters

For serial communications, the two sides must agree on communications data format, otherwise the receiver can not receive data. Meanwhile, a test bit should also be set to improve the accuracy of data transmission. When the error of data transmission occurs, it can be indicated.

The main parameters of communications format settings:

Baud rate: Data are transmitted by way of bit as a unit, so the transmission time of each bit must be set. It is normal to represent how many bits are transmitted per second. 1200 kbit / s, 2400 kbit / s, 4800 kbit / s, 9600 kbit / s, and19200 kbit / s are very common.

The number start bit: The bit which transmits data at the beginning is called start bit. Both sides must agree on the number of the start bit before the communications in order to coordinate. The number of start bit is 1.

The number of data bit: The number of bits that transmits data once is called the number of data bit. When transmitting data, in order to improve the efficiency of data transmission, the transmission each time is not only one bit but several ones. It is usually eight bits, just one byte. Normally seven bits are used to transmit ASCII code.

Test bit: In order to improve transmission reliability, a test bit is generally set to detect whether there are errors in the transmission process. It generally occupies one bit. The way of test commonly used are even test and odd test. Of course, it is available for you not to use test bit.

Even testing requires the number of transmission data and test bit “1” (binary) be an even number. When the number is not even, it means the errors of data transmission.

Odd testing requires the number of transmission data and test bit “1” (binary) be an odd number. When the number is not odd, it means the errors of data transmission.

Stop bit: When the number of a data bit at a time transfer is completed, the signal of transmission completion should be sent, which is called stop bit. Normally stop bit is in the form of 1 stop bit, 1.5 stop bits and 2 stop bits.

Station number: In communications network, in order to mark the different stations, each station must be given an unique representative character which is called as the station number. Station number can also be referred to an address. All of the stations in network can not be the same station number, otherwise there will be communications chaos.

Think about it: Baud rate bit is 9600 kbit /s, 8 data bits, 1 stop bit, 1 even testing bit, 1 start bit. Question: how many bytes can be transmitted at most per second?

4. Introduction of S7-200 Communications Protocol

S7-200 communications interface is defined as shown Fig.2-56. S7-200 connects RS-485 signal B with RS-485 signal A in a communications, and multiple PLCs can compose a network.

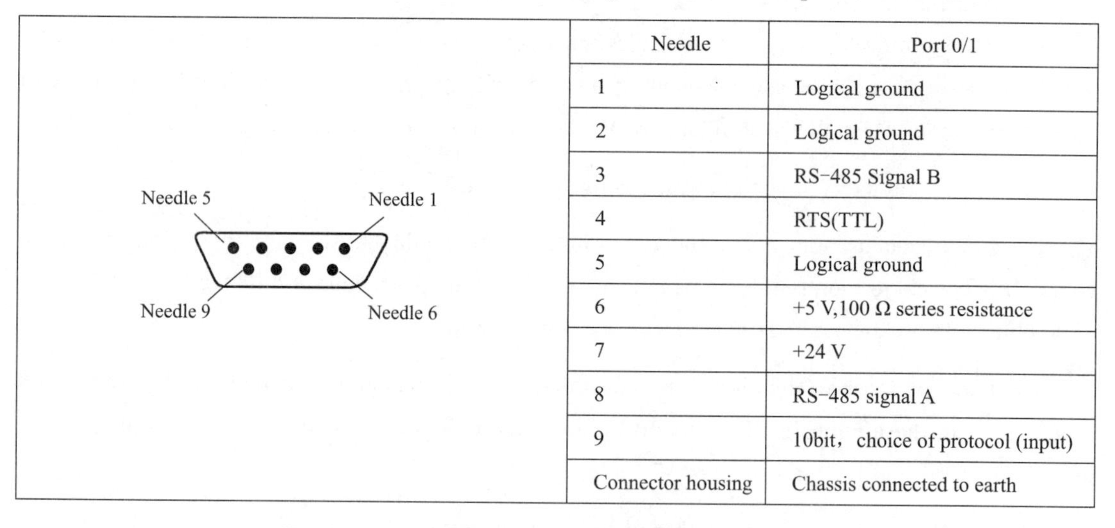

Needle	Port 0/1
1	Logical ground
2	Logical ground
3	RS-485 Signal B
4	RTS(TTL)
5	Logical ground
6	+5 V,100 Ω series resistance
7	+24 V
8	RS-485 signal A
9	10bit，choice of protocol (input)
Connector housing	Chassis connected to earth

Fig.2-56 S7-200 Communications interface definition

S7-200 communications interface is RS-485. Communications protocol is available to use PPI protocol or Modbus protocol. Self-defined communications protocol is used to communicate data through S7-200 communications instructions.

When using PPI protocol to communicate, only one PLC or other device is used as the communications initiator, which is called the master station. And the other PLC or devices can only transmit or receive data passively which are called slave station. Network devices can not send data at the same time, otherwise it will cause the network communications error.

PPI communications protocol format is not introduced here. Only its communications parameters are given: 8 data bits, 1 even testing bit, 1 stop bit, 1 start bit. Communications speed rate and station address can be changed according to actual situation.

Set S7-200 PPI communications parameters

S7-200 acquiesced communications parameters: The address is 2, the baud rate is 9600 kbp /s, 8 data bits, 1 even testing bit, 1 stop bit and 1 start bit.

The address and baud rate can be changed according to actual situation, other data formats can not be changed. To set the PLC communications parameters, select the command of "communications port" of "system blocks". Start to set the address and baud rate when the following window appears, as shown in Fig.2-57.

After setting the parameter, the data must be downloaded to the PLC, and the "system block" option is selected, or set parameters are not effective in the PLC, as shown in Fig.2-58.

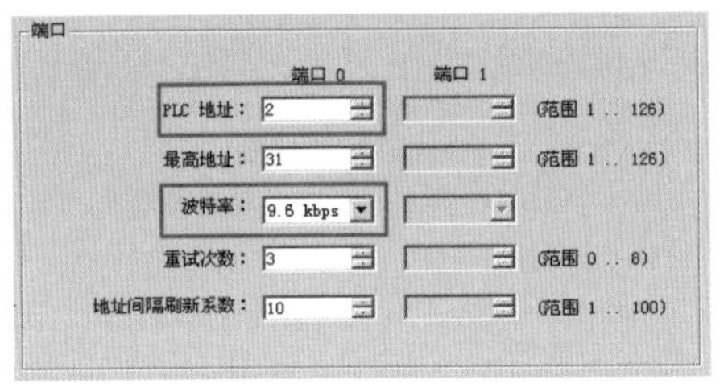

Fig.2-57 PLC address and baud rate setting.

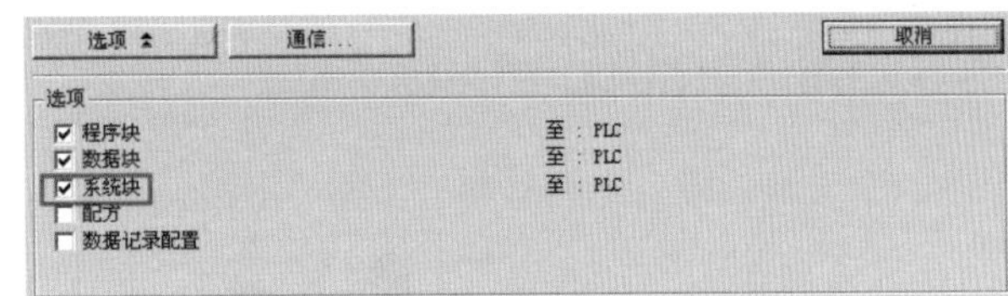

Fig.2-58 Communications data download.

Summary of Knowledge and Skills

Note: Serial communications is commonly used in industrial sites. S7-200 PLC's communications port is physically a RS-485 port. The acquiesced communications protocol is PPI. When the customers use network read and write commands and instruction program, you should notice that the consistency of communications parameters setting between two or more PLCs. There is only one master station in the master-slave mode. PPI is a master-slave protocol communications. In master-slave station, within one command network the master station sends request to the slave station, and the slave responds. The slave station does not send information and just waiting for the requirements of the master station and giving response. If PPI master station mode can be used in user's program, you can read and write information from the slave station by using network read and write commands in master station.

Engineering Competence Training

Referring to S7-200 system manual, think about how to use MODBUS bus, PROFIBUS bus and Ethernet connection with S7-200 PLC.

Development Training — Network read and write command.

Complete the following tasks with the guide of the network read and write command . There are three PLCs in network, address of A is 2; address of B is 10; address of C is 11. A PLC is required to read B PLC VB100 values and C PLC VB102 values once per minute.

With the communications technology I will get everything through.

Task Seven Application of Human-Machine Interface and Configuration in the APL

Task Objectives

1. Grasp the concept and characteristics of human-machine interface, and its configuration method;

2. Being able to write configuration program of the human-machine interactive configuration program and to perform installation and testing.

The function of PLC is powerful and can complete a variety of control tasks. But we can see that it can neither display data nor have a nice interface. It's unlike the computer control system, which could graphically display data and be easily and conveniently operated.

With intelligent terminal device which is human-machine interface, with the configuration software provided from the human-machine interface can easily design the interface required by the user. It can also be operated in human-machine interface.

The human-machine interactive way provided by human-machine interface device is just like a window which is the conversational interface between the operator and the PLC. The status of PLC, the current process data and fault information are graphically displayed by human-machine interface. Users can conveniently operate and observe the monitoring devices or systems by using HMI device. Industrial-touch screen has become one of the indispensable human-machine interface devices in modern industrial control systems. As shown in Fig.2-59 is a number of industrial-touch screens.

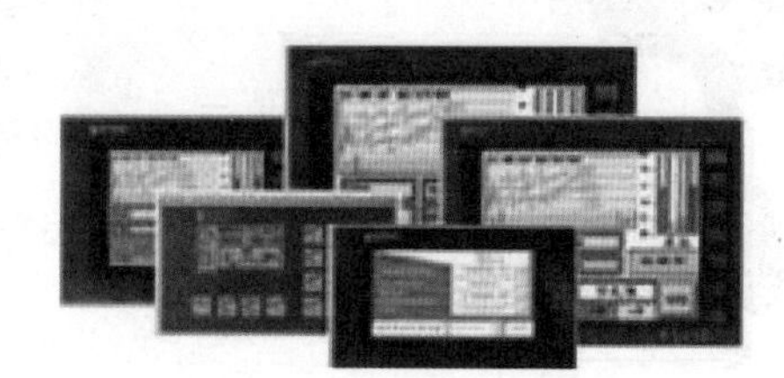

Fig.2-59 Touch screen

YL-335B uses TPC7062K human-machine interface developed by Kunluntongtai.

In YL335–B Automatic Production Line, we can observe, grasp and control the Automatic Production Line and working status of PLC through the window of the touch screen, as shown in Fig.2-60.

Fig.2-60 YL335-B Automatic Production line

Subtask One Getting to Know TPC7062K Human-machine Interface and MCGS Embedded Industrial Control Configuration Software

TPC7062K is a high-performance embedded integrated industrial control machine with low-power consumption and CPU centered. The design of the product uses a 7-inch high brightness TFT LCD screen (resolution ratio 800 × 480), four-line resistance touch screen (resolution ratio 4096 × 4096). Meanwhile it also pre-installed embedded real-time multi-task Microsoft operating system-WinCE.NET (Chinese version) and MCGS embedded configuration software (running version).

MCGS embedded configuration software is developed by Kunluntongtai Company and adopted in mcgsTpc series human-machine interface devices. It can mainly complete field data collection, monitoring, frontier data processing and control.

1. Simple Use of TPC7062K

Fig.2-61 is front and back views of mcgsTpc7062k.

Fig.2-61 Front and back views of mcgsTpc7062k

The power incoming line of TPC7062K main-machine interface and various communications interfaces are carried out on the back.

(1) Interface instructions

Back view of TPC7062K see Fig.2-62

Item	TPC7062K
LAN（RJ45）	Ethernet interface
Serial port (DB9)	1×RS-232，1×RS-485
USB1	Master port, USB1.1 compatibility
USB2	Slave port: Used for download engineering
Power interface	24 V ±48 V DC

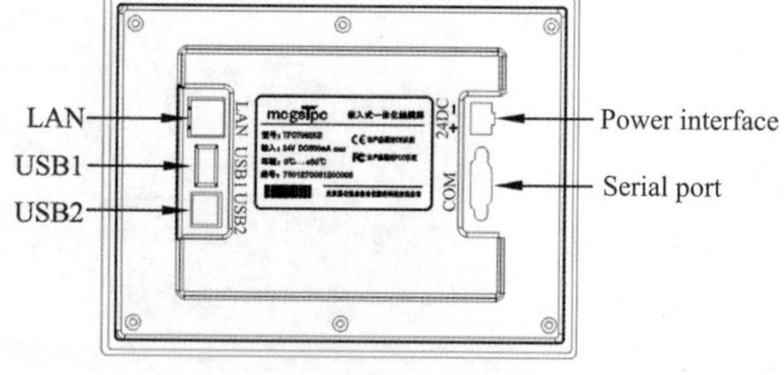

Fig.2-62 Back view of TPC7062K

(2) Pin definition of serial interface

Pin of serial interface see Fig.2-63

Interface	PIN	Pin definition
COM1	2	RS-232 RXD
	3	RS-232 TXD
	5	GND
COM2	7	RS-485 +
	8	RS-485 −

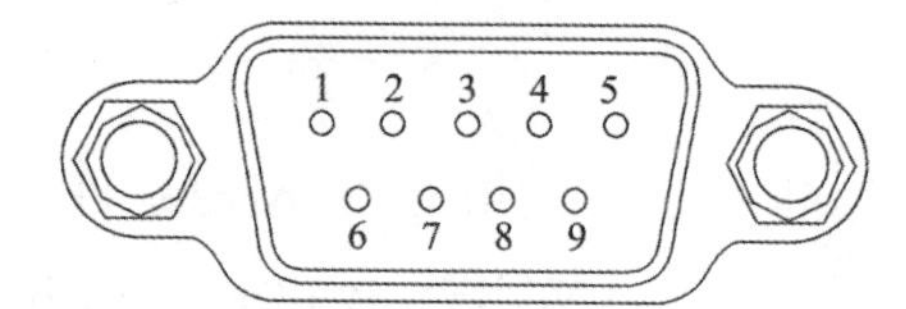

Fig.2-63 Pin of serial interface

(3) Diagram of the power plug and the pin definition

Diagram of the power plug see Fig.2-64

PIN	Definition
1	+
2	−

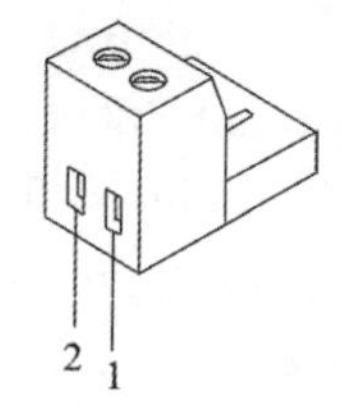

Fig.2-64 Diagram of power plug

(4) Initiation TPC7062K

TPC7062k is powered with 24 V DC power. The “Starting” tip bar will appear on the screen after turning on the machine. At this time the system will automatically enter the running interface without any operating system.Shown as Fig.2-65.

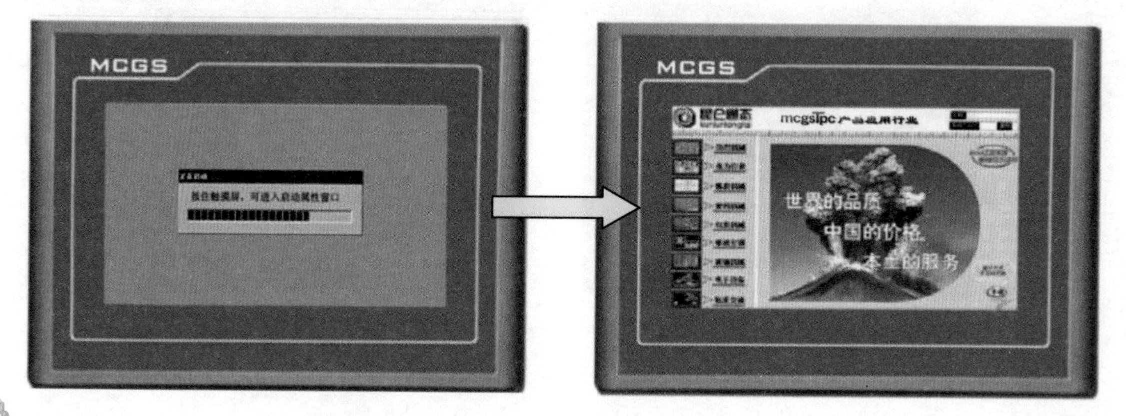

Fig.2-65 TPC7062K initiating and interface running

2. Getting to Know MCGS Embedded Configuration Software

Combining with other related hardwares, MCGS embedded configuration software is capable of quickly and easily developing devices for field acquisition, data processing and control devices. It can configure a variety of special devices such as: intelligent instruments, data acquisition module, paperless recorder, non-attendant field gathering station, human-machine interface, etc.

(1) Main functions of MCGS embedded configuration software

① Simple and flexible visual operating interface with all-Chinese and visual development interface meet Chinese habits and requirements.

② It is a real 32-bit system with strong real-time and better parallel processing properties and parallel processing tasks in the unit of line stroke.

③ Plentiful and vivid media images: providing the operators relevant information in the way of images, icons, report forms, curves and so on.

④ Perfect security system: providing good security system and setting different operating limits of authority for users of different levels.

⑤ Strong network function: powerful network communications function.

⑥ Variety of alarm functions: providing different alarm modes, being with various alarm types, convenience of setting alarm for users.

⑦ Supporting a variety of hardware devices.

In short, MCGS embedded configuration software has the same great functions as that of the general configuration software. It is easy to operate, learn and use.

(2) Composition of MCGS embedded configuration software

MCGS embedded user's application system is composed of five-parts: main control window, device window, user window, real-time data base and running strategy, as shown in Fig.2-66.

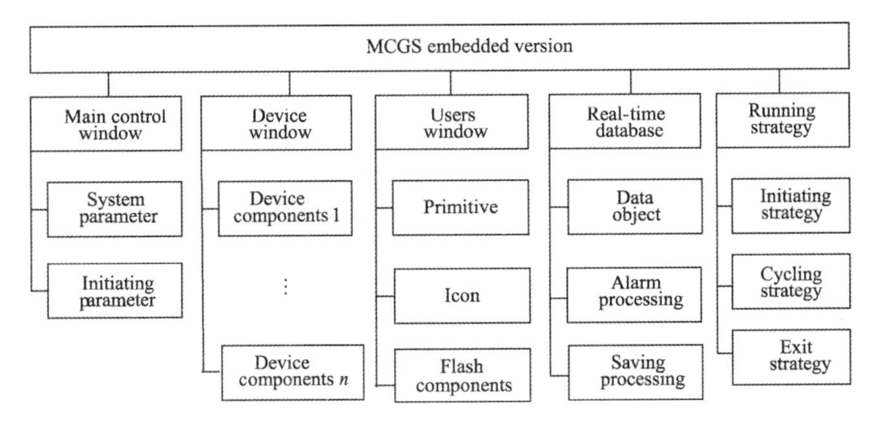

Fig.2-66 Composition map of MCGS embedded configuration software

Main control window structures the main frame of application system. It is the main frame which determines layout of the engineering operation in industrial control, running process, characteristic parameter, startup characteristic, etc. Device window is the media for connecting the MCGS embedded version system with the external device. Device window is designed to store components for different types and functions equipment or device in order to operate and control external devices. The device window also collects the data of external device through components and then input them into the real-time data base or output the data of real-time data base to the external device. User window, which achieves the "visual" data and process, can be placed in three different types of graphic objects: primitives, icons and flash components. Users can construct a variety of complex graphical interfaces and achieve the visualization of data and process in different ways through placing different graphical objects in users' window. Real time data base, which is the core of MCGS embedded system, works as a data processing center and also acts as a common data exchange area. Real-time data collected from external device is input to the real-time data base. Operating data from other parts of the system also comes from real-time data base. Running strategy, a frame provided by the system, is an effective control means in system's running process. Running strategy is placed with strategy line composed of strategy condition components and strategy components. It enables the system to perform the tasks in accordance with the set order and conditions and realizes the precise control of work process for external device through defining the running strategies.

(3) System structure of embedded system

The configuration environment and the simulated running environment of embedded configuration software are equivalent to a complete set of tools software, which can run on PC machine. Embedded configuration software's running environment, being an independent operating system, performs various processing according to the way specified by users in configuration engineering and complete the goals and functions designed by the user configuration. Operating environment itself does not have any meaning. It must work together with the configuration engineering as a whole and constitute the user's application system. After the completion of configuration work, the configured engineering can be downloaded to operating environment of the integrated embedded touch screen through the USB port. Then the configuration engineering can leave the configuration environment and independently work on TPC. Thus, It achieves the reliability, real time, certainty and security of the control system. The connection of TPC7062K with configuration computer is shown as Fig.2-67.

Fig.2-67 Connection of TPC7062K with configuration computer

Plug flat end of common USB into the computer's USB port, the other micro end into the USB2 port of the TPC side.

Subtask Two Wiring of TPC7062K and PLC and Engineering Configuration

1. Wiring of TPC7062K with PLC

After understanding TPC7062K, then we will learn the communications way of it with Siemens S7-200PLC. The way of wiring is shown as Fig.2-68.

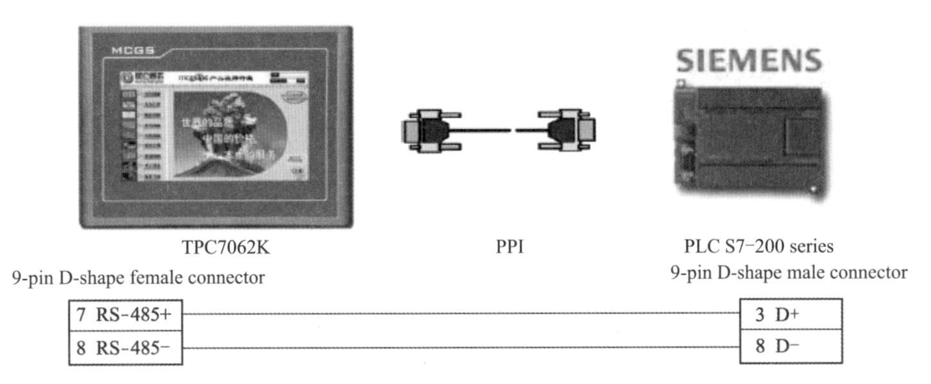

Fig.2-68 Connection between TPC7062K and Siemens S7-200 PLC

After MCGS embedded configuration software is installed in computer, the desktop adds the following two shortcuts icons, as shown in Fig.2-69, which were used to start.

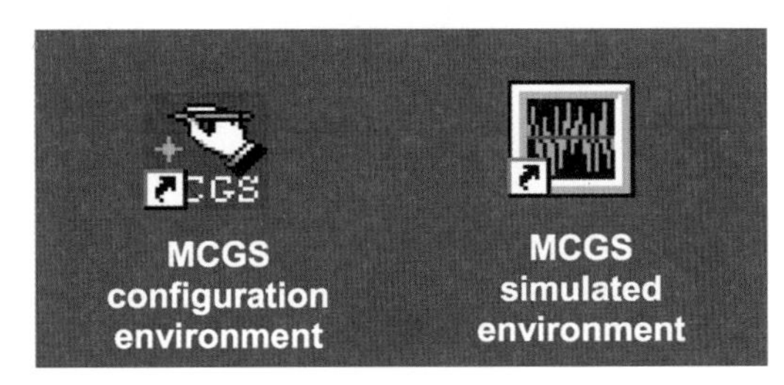

Fig.2-69 Icons of MCGS embedded configuration environment and simulated running environment

2. Configuration of Connecting MCGS Embedded Version with Siemens S7-200 PLC

Next, we will briefly introduce the configuration process of connecting MCGS embedded version with Siemens S7-200 series PLC. Let's hand-on now!

(1) Project establishment

Double click the shortcut of configuration environment on Windows operating system desktop to start the embedded configuration software. Then to establish communications engineering as the following steps:

① Select the "New Project" option in file menu, then pop-up the dialog box(See Fig.2-70), the section of the TPC type is "TPC7062K", click OK button.

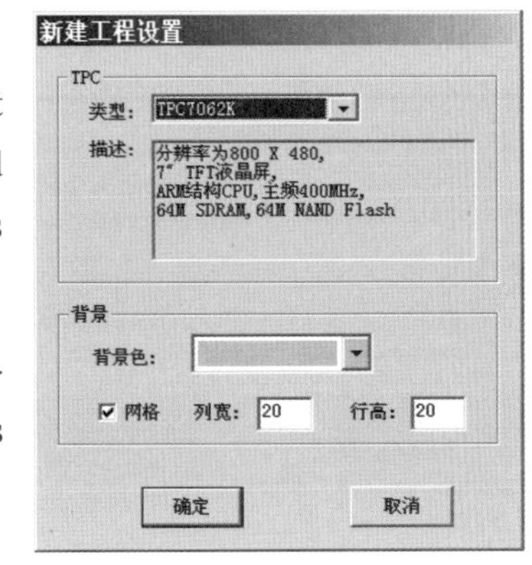

Fig.2-70 Types of choosing human-machine interface

② Select the "Save Project As" in the File menu, then pop-up window - save the file.

③ Enter "TPC communications control engineering" in the File Name Column, click "Save" button, then engineering has been created.

(2) Engineering configuration

Next we will verify the communications link correctness between touchscreen TPC7062K and Siemens S7-200 by writing real example program. Please refer to the manuals of specific action steps in the disk provided.

Development Training—Design Configuration Program of YL-335B Production Process Monitoring

To meet the following functions: ①User management; ②Entire production line monitoring, status and flash procedures displaying ③Individual station status checking; ④Parameter setting; ⑤Production statistic; ⑥Historical record of failure warning. Students can score according to Table 2-4.

Table 2-4 Reference marking table of skills assessment

Name		Group		Start time		
Specialty/Class				End time		
Project content	Test requirement	Grade	Grading criterion	Deduct	Self evaluation	Mutual evaluation
1. User management	Three levels 's users: checking, operating and parameter revising	15	Meet the function of user password authentication and grading operation, 5 points will be deducted if one function is failure.			
2. Monitoring the entire production line, displaying status and animation process	Picture shown clearly and displaying correctly data status	30	Beautiful human-machine interface get 10 points, correct of status data get 10 points, animation process get10 points			
3. Individual station status checking	Picture shown clearly and displaying correctly data status	20	4 points for each station			
4. Parameter setting	Writing and setting the parameter of frequency converter and picture of number of workpiece .	10	Being able to realize correct parameter setting.			
5. Production statistic	Writing the picture of current production status of the production line	10	Being able to obtain correctly the statistical information of workpiece in production line			
6. Historical record of failure warning	Writing the picture of abnormal warning records of the production line	5	Being able to alarm and make records when failure and abnormal cases appear in the production line.			
7. Professional quality and team work spirit	Team work cooperation, clear division of task, detailed technical materials	10	Referring the working records, program designing manual.			
Teacher's feedback		Mark (teacher)	Total mark			

 Summary of Knowledge and Skills

 Note: With the extensive uses of touch screen in industry, the human-machine interface and configuration technology achieve the human-machine visual interaction. The products of human-machine interface consist of hardware and software. Hardware includes processors, display unit, input unit, communications interface, data storage unit, etc. The processor's performance determines the high or low property of HMI products, which is also the core of HMI unit. The human-machine interface based on touch screen consists of touch screen, touch screen controller, microprocessor and relevant software. HMI software is generally divided into two parts, namely, system software running in the HMI hardware and the picture configuration software running on computer's Windows operating system. The programming of configuration software is simply and easy to do maintenance. Human-machine interface system can display working status of devices or equipment, such as indicator light, buttons, words, graphics, curve, data, words input operation, printing output, production formula storage, real time and historical data records of equipment during the operation, simple logic and numerical calculation. Meanwhile it can be connected to a variety of industrial control devices networking

 Engineering Competence Training

Search MCGS human-machine system manual, write the configuration program, design user rights limit management, production statistics, history curve, failure alarm records and functions interface.

Human-machine interface and configuration technology can link unboundedly and it can also make what is good still better !

Installation & Testing of Automatic Production Line

Chapter Three

Project Acceptance —Installation and Testing of Units in the APL

The battle is about to begin. I'm not sure what to do!

Don't worry! So long as you really master what you have learned: 5 routines and we are sure to win. Now let's begin!

Through the study of the core technology in Chapter Two "project preparation", we have mastered knowledge points needed for installation and testing of the Automatic Production Line(APL). Now, with the APL YL-335B as an example, we are to meet a challenge of the skills, that is to say we'll do the training of testing for installation and program design at each substation.

YL-335B Automatic Production Line comprises 5 stations. Each of them can run as an independent unit or can be combined with any others, which represents the application of PLC in different working environment, different applied field, different applied aging situation. Each station has a main technical unit, together with some other technical units. Reuse of PLC technology in different situations indicates its key position in the electro-mechanical control field. It represents course development concept of integrated teaching environment and PLC core technology.Its schematic diagram shown as Fig. 3-1.

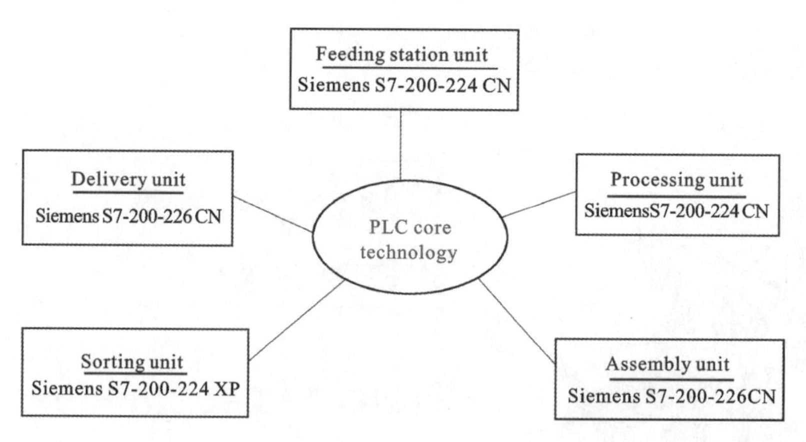

Fig. 3-1 Integration of PLC core technology with teaching environment

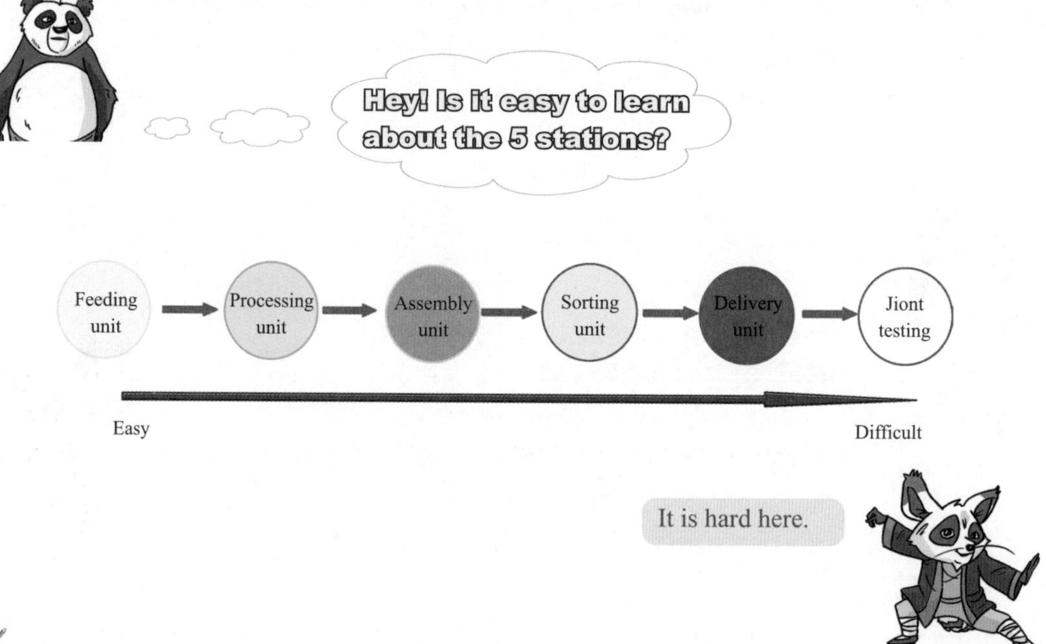

Task Objectives

1. Being able to conduct installation and testing for each station in the Automatic Production Line;

2. Being able to do the control programming and testing for each station in accordance with the control requirements;

3. Being able to solve problems during installation and operation of the Automatic Production Line.

Training Mode

Three students in one group cooperate with one another to complete the installation and testing of the 5 the stations in the Automatic Production Line. To achieve the goal of the training, let's start with the feeding unit.

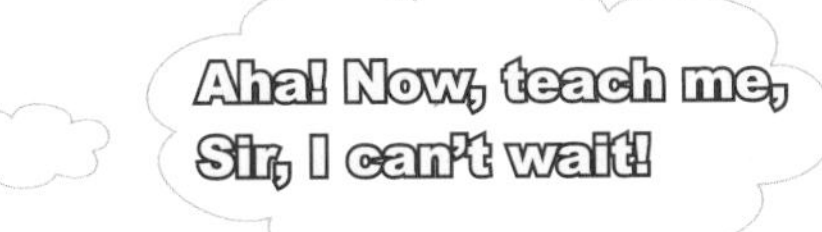

Task One Installation and Testing of the Feeding Unit

Ok! Let's start with the routines of the feeding unit

Subtask Getting to Know the Feeding Unit

The feeding unit (See Fig. 3-2) is the starting unit in the Automatic Production Line, which supplies materials to other units in the system. It is equivalent to an automatic feeding system for a real production line. Its main structures are feeding pipes for workpieces, workpiece pusher device, support, valves set, terminal board assembly, PLC, emergency stop button and start/stop button, cable trough, base plate, etc.

Fig. 3-2 Feeding Unit

Functions of the Feeding Unit

The function of feeding unit is pushing workpieces (raw material to be processed) in the feeding bin automatically out onto a charging table as needed, so that a manipulator on the delivery unit will pick it and deliver it to other units. See Fig. 3-3 real object of the feeding unit.

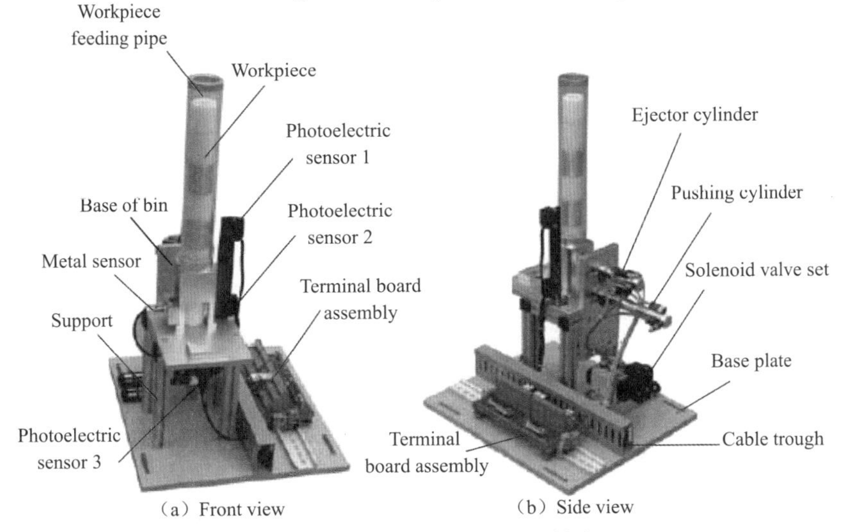

Fig. 3-3 Overview of the Feeding Unit

Operation Process of the Feeding Unit

Workpieces are vertically piled in the feeding bin. The pushing cylinder is located at the bottom of the feeding bin and its piston rod can move through the bottom of the feeding bin. When the piston rod is back at home position, it is at the same horizontal level as the lowest workpiece, and the clamping cylinder is at the same horizontal level as the sub-lowest workpiece. When workpieces are needed to be pushed onto the material table, the piston rod of the clamping cylinder is first to be pushed out and to hold sub-lowest workpiece tightly. Then the piston rod of the pushing cylinder is pushed out and to push the lowest workpiece onto the charging table. After the pushing cylinder gets back and pulls out from the bottom of the feeding bin, the clamping cylinder can then returns and releases the sub-lowest workpiece. In this way, the workpieces in the feeding bin will move downward by gravity another one ready to be pushed out. See Fig. 3-4 shows the main structure of the feeding unit.

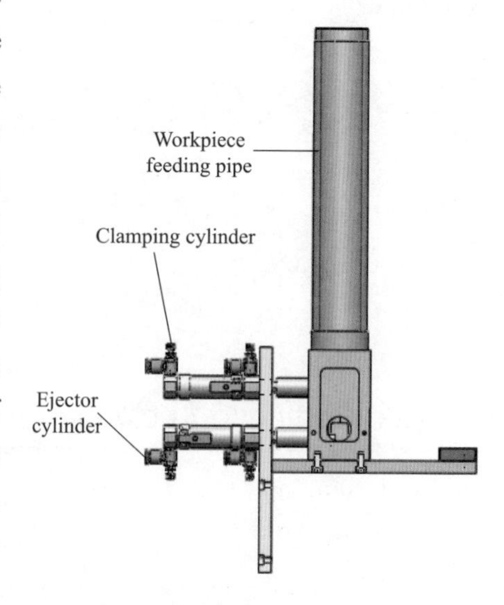

Fig. 3-4 Main structure of the feeding unit

A diffusion photoelectric proximity switch is installed at the fourth layer of workpiece at the base and the feeding pipe. They are designed to detect whether there are sufficient material in the feeding bin.

If there is no workpiece, both of the two diffusion photoelectric proximity switches at the bottom and the 4th layer are in normal state. If there are only 3 workpieces starting from the bottom, then the switch at the bottom is actuated while the switch at the 4th layer remains in normal condition. It indicates that the workpieces are running out. So the two photoelectric proximity switches will give signal whether there are sufficient or insufficient workpiece in the feeding bin.

The pushing cylinder pushes the workpiece onto the material table. There is a hole on the table. Under the table there is a cylindrical diffusion photoelectric proximity switch that sends light upward during operation. In this way, the detect of whether there being the workpieces can be conducted through the hole and a signal will be sent to the system allowing it to know whether or not there are any workpieces on the table. The signal can be used in the control program of the delivery unit to determine if it is necessary to start the manipulator to pick out the workpiece.

Knowledge Development

1. How do you reflect it in the program whenever there are less than 4 workpieces in the feeding bin and the sensor gives an alarm?

2. How do you realize the transfer between a manual single cyclic control, a manual single step control and automatic control in the programming?

3. How do you reflect in the configuration interface the number of completed workpieces and the number of metal material?

Summary of Knowledge and Skills

Through training, you have got a good knowledge of the structure of the feeding unit; in personal practice, understand the application of pneumatic control, sensor and PLC control technologies. And you know the combination of them in one unit which represents the practical application of electro-mechanical integration.

Engineering Competence Training

Master the way of engineering and develop a rigorous working style.

Task Two Installation and Testing of the Processing Unit

Fig. 3-5 Overview of the processing unit

Key points of task two: Design the pneumatic and control circuit according to functions of the processing units, then follow the correct steps to conduct installing and testing.

Routine two is very interesting;
I'm going to work hard!

Subtask Getting to Know the Processing Unit

The function of the processing unit is to take workpieces to be processed from the material table to a place right below the stamping cylinder for stamping in the processing area, and then bring the workpieces back to the material table. Fig. 3-6 shows the processing unit.

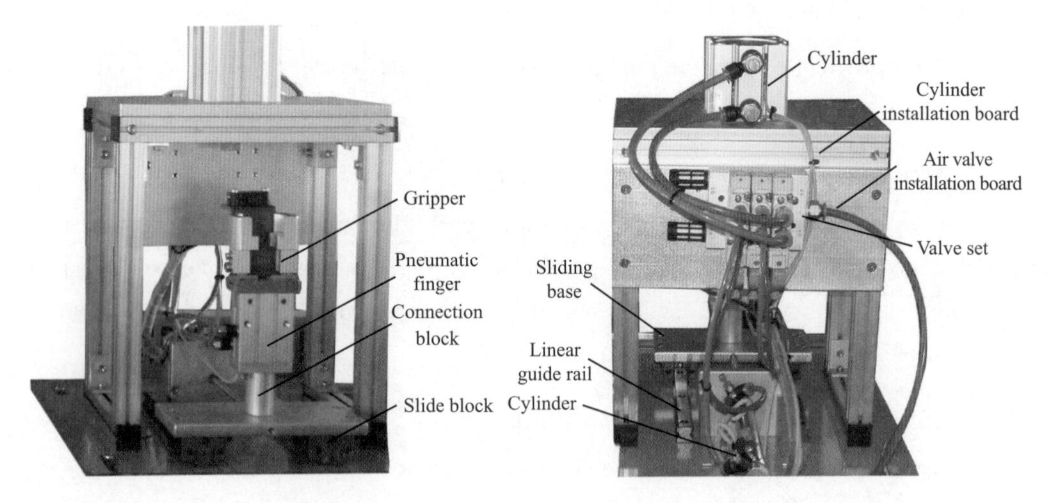

Fig.3-6 Processing unit

The material table is used for fasten a workpiece to be processed and take the workpiece to the place right below the processing (stamping) unit for stamping. It mainly consists of clamp, pneumatic fingers, extension and retracting cylinder, linear guide rail and its sliding block, magnetic inductive proximity switch, diffusion photoelectric sensor. and sliding device, shown as Fig. 3-7.

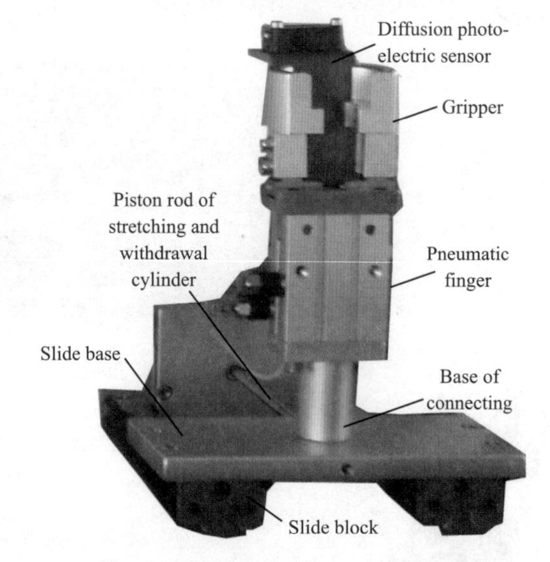

Fig. 3-7 Material table and sliding device

Following normal working of the system, initial status of the sliding material table is that the stretching and withdrawal cylinder extends and the pneumatic finger of the material table opens. The processing procedure is as the delivery unit has delivered the material onto the material table and the sensor has detected it, PLC control program then instructs the pneumatic finger to clamp the workpiece→the material table moving back to the place right below the stamping cylinder in the processing area→piston rod of the stamping cylinder stretching down for stamping the workpiece→retract upward after the action of stamping→material table reaches out again→when in place, the pneumatic fingers loosen, and the processing is complete. Then the signal of completion is sent to the system; the preparation is made for the next processing.

There is a diffusion photoelectric switch on the moving material table. When the material table is empty, the diffusion photoelectric switch will be in normal state. When there are workpieces on the material table, the switch will come into action, indicating there are workpieces. The photoelectric

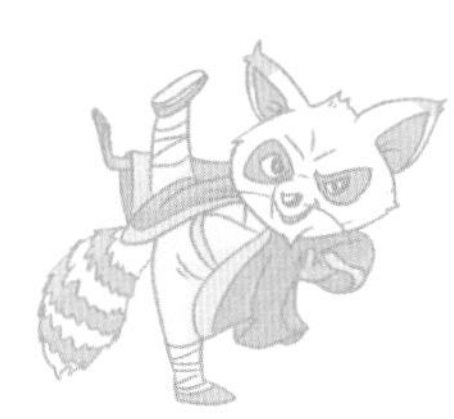

sensor sends signals to the input end of PLC at the processing unit, which are used to determine if there are any workpieces to be processed on the material table. The material table will return to its initial place after the processing completed.

Diffusion photoelectric switches installed on the sliding material table are still the type of photoelectric switch (small beam type) with CX-441 built-in amplifiers. Positioning for the table stretching and withdrawal to its place is completed through adjusting the two magnetic switches on the stretching and withdrawal cylinder. The required withdrawal position is the place right below the processing ram and the stretching position to such an extent that it fits the manipulator of the delivery (transferring) unit as a whole, in order that the manipulator is able to smoothly deliver the workpiece to be processed to the material table.

The processing unit is used for workpiece stamping. It mainly consists of stamping cylinder, stamping ram, installation plate, etc. The processing (stamping) unit is shown as Fig. 3-8.

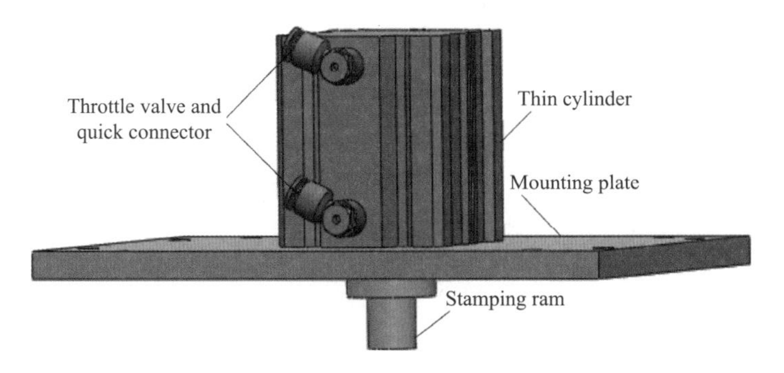

Fig. 3-8 Processing(stamping) Device

When a workpiece reaches the stamping position and the extension cylinder piston rod retracts in place, the stamping cylinder extends to process the workpiece. As soon as the processing is over, the stamping cylinder moves back to get prepared for the next stamping. The ram, which is located at the head of the stamping cylinder, performs the stamping procedure in accordance with requirements for the workpiece. The mounting plate is used for install and fix the stamping cylinder.

Wow! The Assembly
unit is complicated.
I'll work hard!

Task Three Installation and Testing of the Assembly Unit

The assembly unit is capable of simulating two material assembling processes, and can also simulate material transferring process through rotary worktable. Fig. 3-9 is a real object of an assembly unit

Fig. 3-9 Real object of assembly unit

Now, let's start to
learn the routines of
the assembly unit.

Pay attention to these when learning the routines: Design pneumatic and control circuit according to the functions of the assembly unit and follow the correct steps to conduct installation and testing.

Subtask Getting to Know the Assembly Unit

The assembly unit is designed to assemble two different sizes, small cylindrical workpieces together, i.e. taking small cylindrical workpieces out of the feeding bin (black or white, See Figure 3-10), into a center hole on the material table.

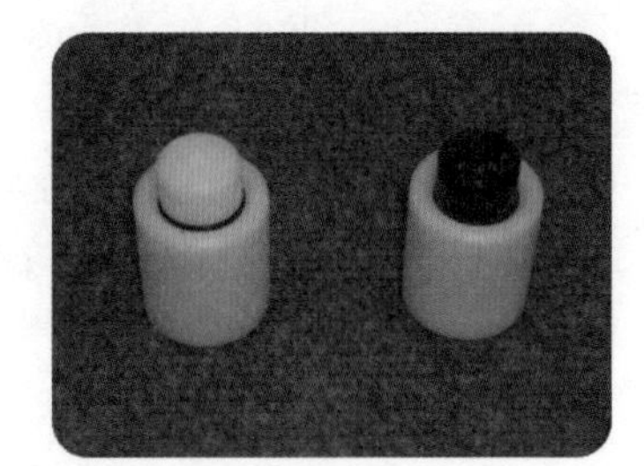

Fig. 3-10 Small cylindrical workpieces

Let me see, what
weapon shall I need
to complete...

An overall view of the assembly unit is shown as Fig. 3-11. Basic working process of the assembly unit is as following. Material from the feeding bin automatically fall down by force of gravity. Two ejecting and pushing linear cylinders are used to grip and release material at the bottom layer and to distribute falling down material. Material which is at the lowest bottom falls down to the feeding disk according to the specified routes. After the swing table's position changing of 180° , the manipulator, which is composed of stretching and withdrawal cylinder, lifting cylinder and pneumatic finger, is to do the clamping and displacement and then inserts into a semi-finished workpiece being positioned on the assembly table.

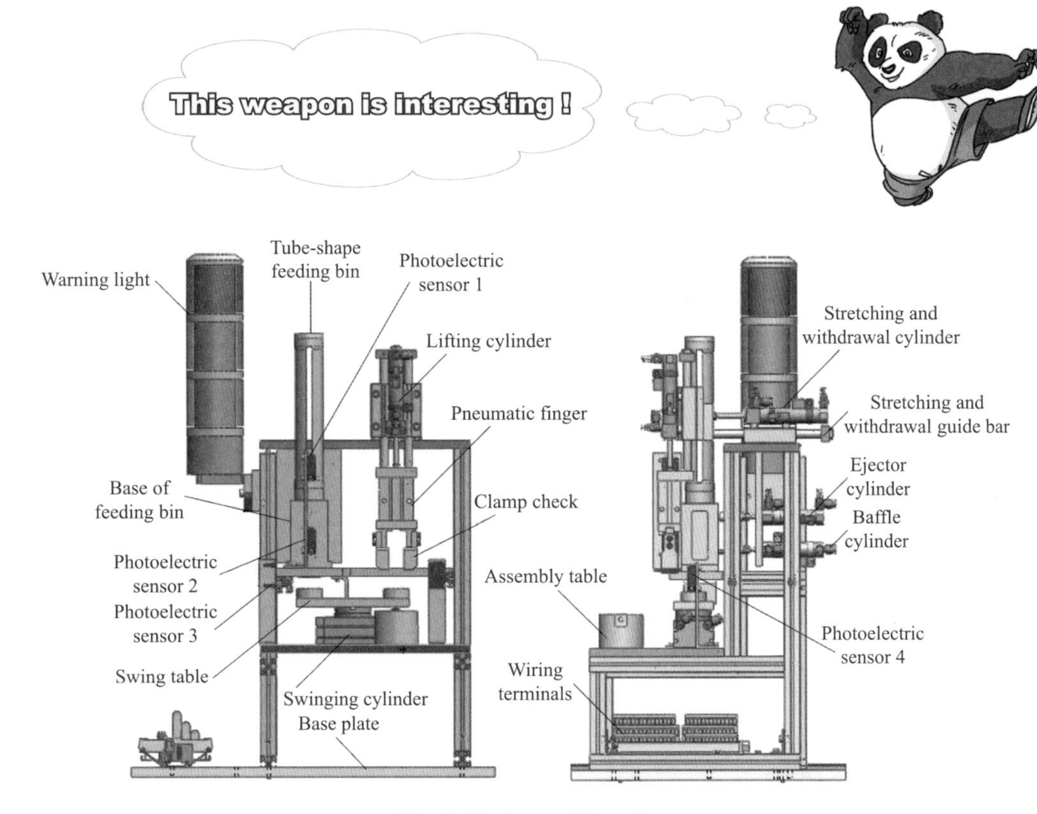

Fig. 3-11 Assembly unit

Structure

The assembly unit mainly consists of simple feeding bins, feeding devices, swing material tables, manipulator, locating devices for semi-products, a pneumatic system and its valve sets, a signal collector and its auto-control system, components of terminals board for electric connection, and in addition, status signal indication lights for the whole production line, aluminum supports and bases for installing other devices, supports for installing sensors and other auxiliary devices.

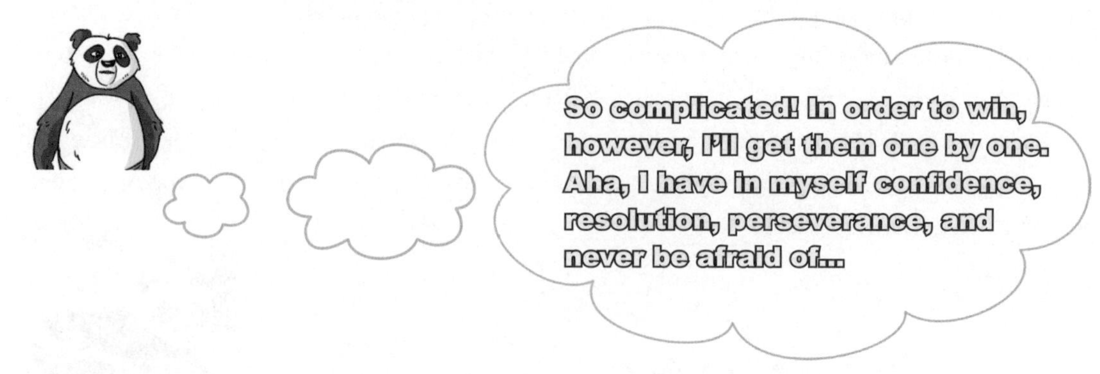

Simple feeding bin

Simple feeding bin is made from a round plastic rod, its real object and diagram shown as Fig. 3-12. It is directly inserted into a connecting hole leading to the feeding unit. A reinforcing metal ring placed on the top can help to protect the hollow plastic column from damaging. Materials are vertically dropped into the column. As there is a clearance gap between the two, the material can fall freely due to gravity.

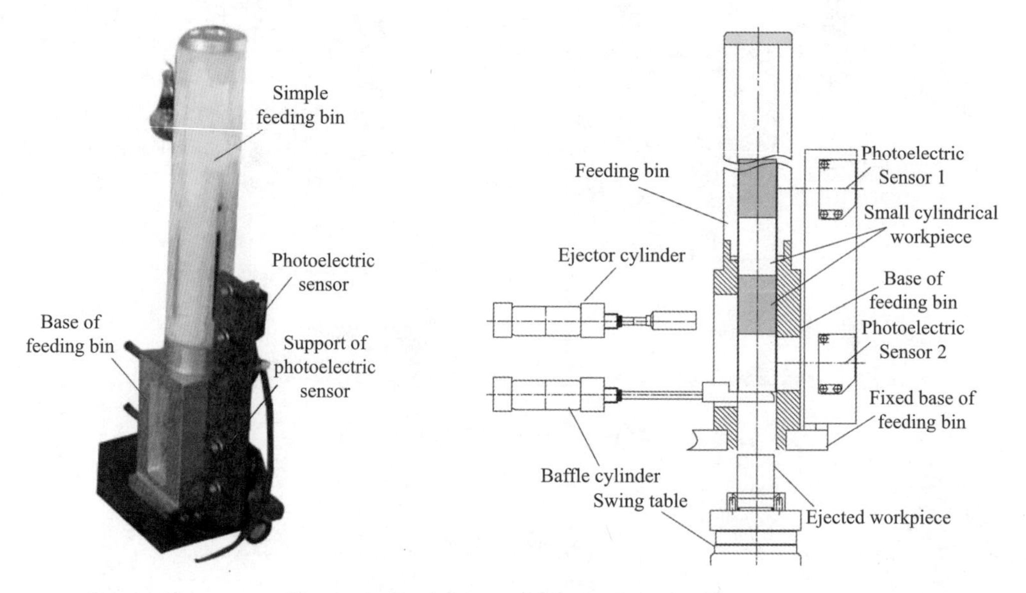

Fig. 3-12 Real object of Simple Feeding Bin

In order to immediately alarm when the feeding bin is short of material, a diffusion photoelectric sensor (CX-441) is installed outside the feeding bin. And plastic column of the feeding bin is with longitudinal grooves so that the infrared ray can smoothly irradiate the inspected materials. Materials in the feeding bin look the same in shape, but different in color, black or white. Detecting the black material is the criterion for sensitivity adjustment of the photoelectric sensor.

Feeding device

The movement of the feeding device is performed by two linear cylinders which is controlled by PLC. The two cylinders are installed up and down and are horizontal movement. The initial location is that the piston rod of upper cylinder withdraws, while the piston rod of the lower cylinder stretches out. In this way, falling materials by gravity are baffled, so this kind of cylinder is called baffle cylinder. As the system is powered on and gets into normal operation, and as soon as the photoelectric sensor at the swing material table detects the needed materials, the piston rod of the upper cylinder, acted upon by solenoid valve, stretches out and baffles the material at the sub-bottom layer, keeping it from falling, so the upper cylinder is also called ejector cylinder.At the same time, the cylinder piston rod of baffle cylinder withdraws, materials fall into the feeding plate of swing table. Then the baffle cylinder resets and stretches out. The ejector cylinder withdraws and materials at the sub-bottom layer fall down and get prepared for the next distribution. Both of the linear cylinders are equipped with magnetic switches for detecting whether the cylinder piston rod has stretches or withdraws. When the system is in normal operation and detects piston magnet steel, the red indicator of the magnetic switch is on and the signal is sent to PLC of the control system.

Swing table

This device consists of a swing cylinder and a material plate. See Fig.3-13. The swinging cylinder drives the material plate to swing up to 180° and sends the signal to PLC through the magnetic switch. The orderly reciprocating motion is achieved under the control of PLC.

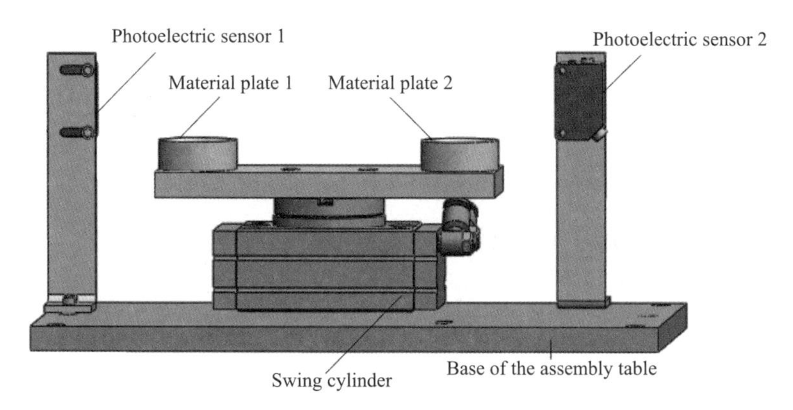

Fig. 3-13 Structure of a swing table

An important component of the swing material table is its pneumatic swinging table. Turning of the swing table is achieved by way of the linear cylinder driving the gear rack. Swing degree is freely adjustable within the ranges of 0～90° and 0～180° degrees. In addition, magnet switches can be installed thereon to detect the signal of turning in place. It is mostly used in devices that require changes in direction and location. See Fig. 3-14.

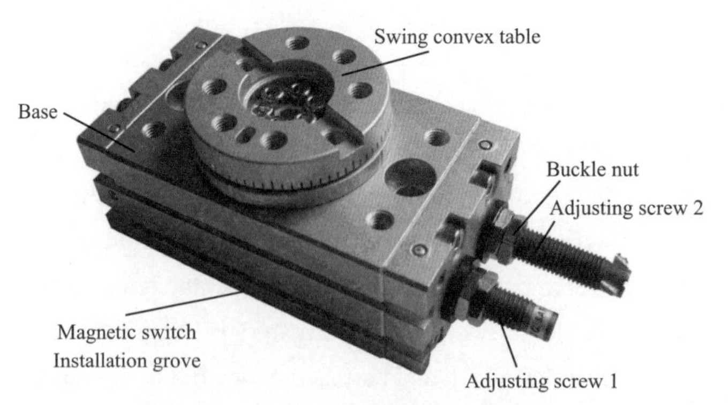

Fig. 3-14 Pneumatic swinging platform

Swing degree of the pneumatic swinging platform used at this station is freely adjustable from 0 ° to 180°. When it is necessary to adjust the rotary degree or precision of swinging position, firstly unscrew the Buckle nut on the Adjusting screw, then turn in or turn out the adjusting screw to change the rotation degree of swing convex table. Adjusting screw 1 and adjusting screw 2 are used for adjustment of left and right handed degree respectively. When the swinging degree adjustment is finished, the reversing nut and base should be locked tightly in order to prevent the adjusting screw from loosening and thus reducing rotation precision.

The signal indicating swing in place is realized by adjusting the positions of the two magnetic switches in the sliding rails of the pneumatic swinging table. Fig. 3-15 shows the schematic diagram of adjusting the position of the magnetic switch. Magnetic switch is installed in the sliding rails of the cylinder. The magnetic switch is available of moving along the sliding rails left and right when releasing the screws of the magnetic switch. When the switch is positioned, the fastening screws can be tightened, and the adjustment of the position is finished.

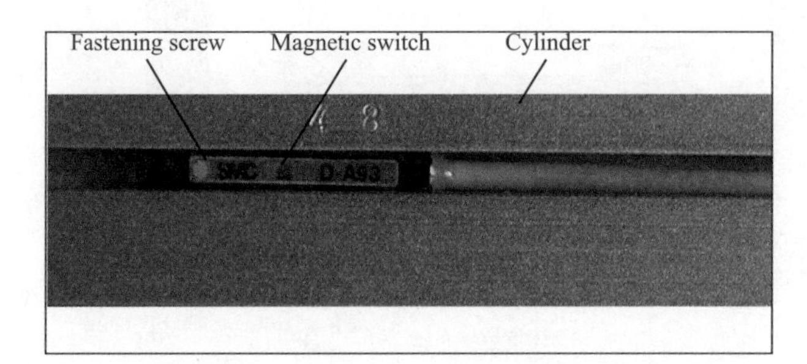

Fig. 3-15 The schematic diagram adjusting the position of magnetic switch

Assembly manipulator

Manipulator is the core of the whole assembly unit. When there is a workpiece on the swing table right below the manipulator, and the workpiece is detected by the positioning sensor of the semi-finished workpiece, the manipulator will start assembly operation from the initial state. The overall shape of the assembly manipulator is shown as Fig. 3-16.

The assembly manipulator device is of a three-dimensional movement device, consisting of two

guide rod cylinders and pneumatic fingers that moves in horizontal and vertical directions respectively.

Overall view of the guide rod cylinder is shown as Fig. 3-17. The cylinder consists of a linear movement cylinder with double guide rods and other accessories.

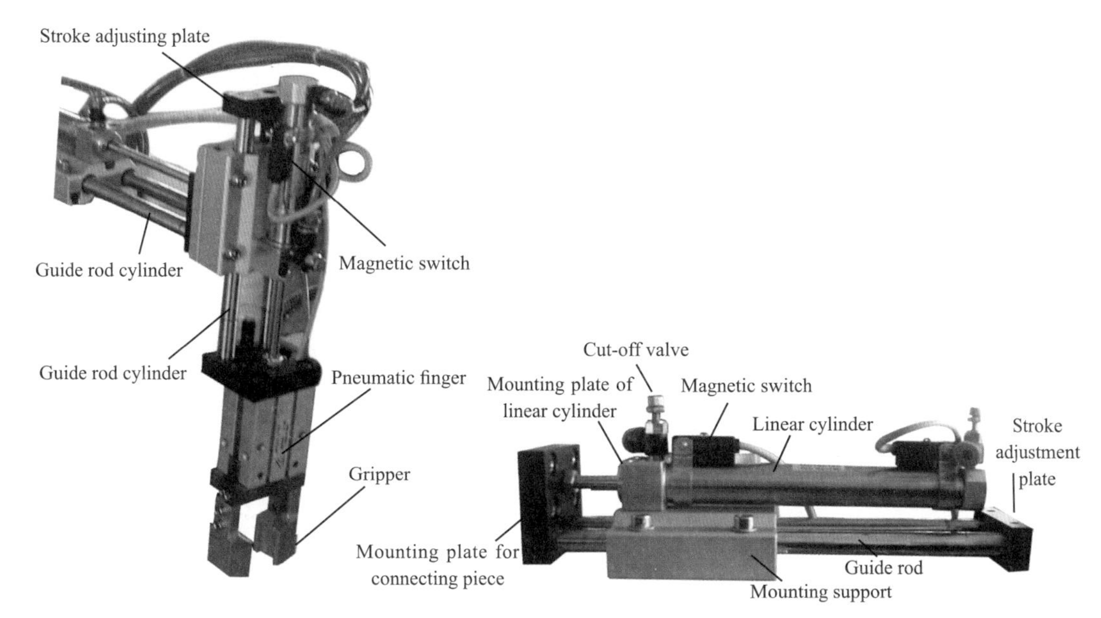

Fig.3-16 Overall view of the assembly manipulator

Fig. 3-17 Guide rod cylinder

The mounting support is used for the installation of the guiding components of the guide rod and the fastening of the whole guide rod cylinder. The connector mounting plate is used for the fastening of other parts to be connected to the guide rod cylinder, and fixing the relative positions of the two guide rods and the linear cylinder piston rod. As one end of the linear cylinder is switched on with compressed air, the piston is driven to move in a straight line together with the piston rod. The two guide rods fixed together by the connector mounting plate stretches and withdraws with the piston rod. In this way the function of the guide rod cylinder is performed. The stroke adjusting plate installed on the end of the guide rod is used for adjusting stretching stroke of the guide rod cylinder. The way of adjusting is to release the fastening screws on the stroke adjustment plate and then allows the stroke adjustment plate to move along the guide rod. As the desired stretching distance is reached, the fastening screws should be completely locked, and the adjustment of the stroke is finished.

Working process of the assembly manipulator: PLC drives an electromagnetic reversing valve that is connected to a vertical movable cylinder. The vertical movement changes to down movement which is conducted by a cylinder with guide rod that drives pneumatic fingers. After moving to the desired position, pneumatic fingers drives the gripper to clamp the materials and sends a signal to PLC through the magnetic switch. Under the control of PLC, the vertical moving cylinder resets to its home position, and the griped materials are lifted up with the pneumatic fingers. As the feeding disc of the swing material table reaches the highest position, the horizontal moving cylinder, driven by the reversing valve, stretches the piston rod and moves to the front side of the cylinder. Then the vertical moving cylinder is driven to move downward again to the lowest position, where the pneumatic fingers are released. After a short time delay, the vertical moving cylinder, the horizontal moving cylinder withdraws and the manipulator returns to its initial state.

During the whole movement of the manipulator, no sensor is designed to detect releasing in place of the pneumatic fingers, while all the other signals of movement in place are collected through the magnetic switches equipped with the cylinder and fed into PLC. Then PLC sends signals to reverse the electromagnetic valves, enabling the manipulator consisting of a cylinder and pneumatic fingers to operate in accordance with a given program.

Positioning Device for Semi-finished Workpiece(Material Table)

A semi-finished workpiece delivered from the delivery unit is directly placed in the material positioning hole of this unit. Positioning is achieved by using a smaller gap between the positioning hole and the workpiece, which helps to complete exact assembling motion and precise positioning. See Fig. 3-18.

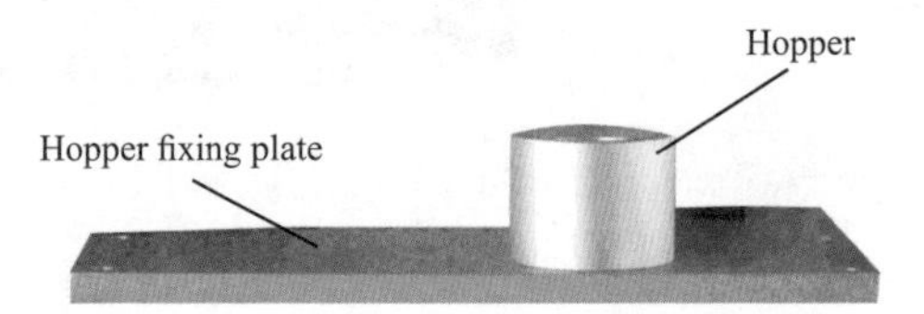

Fig. 3-18 Positioning device for semi-finished workpiece (material table)

Electromagnetic Valve Set

An assembly valve set consists of six 2-position 5-way single electromagnetic reversing valves. as shown in Fig. 3-19. These valves control material distribution, location changing and air circuit of the assembly operation respectively in order to change each of their motion status.

Fig. 3-19 Valve set for assembly unit

Summary of Knowledge and Skills

By training, the students are getting familiar with the structure of the assembly unit, personally practicing and understanding applications of pneumatic manipulator, rotating cylinder control technology, sensor technology and PLC control technology, which are organically integrated in one unit, and also experiencing actual examples in specific applications of electro-mechanical control technology.

Engineering Competence Training

Master the way of the work in engineering and develop a rigorous working style.

Task Four Installation and Testing of the Sorting Unit

I have learned three routines including the feeding unit, the processing unit and the assembly unit. What is the fourth one?

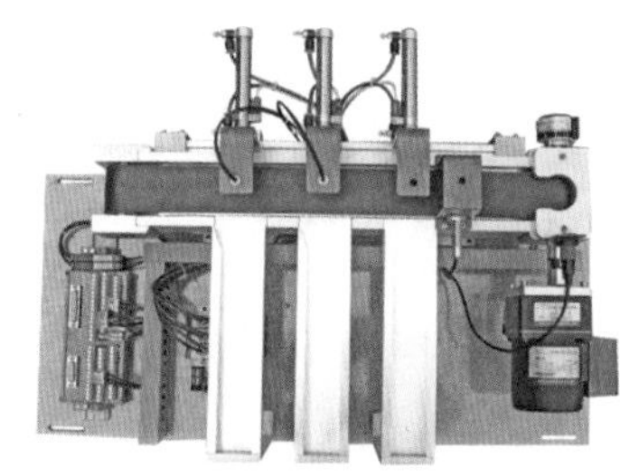

This is the fourth routine: the sorting unit. Look at the Fig.3-20!

Fig. 3-20 The sorting unit

In this unit, you should master certain skills: designing the pneumatic and control circuit according to the function of the sorting unit; following the correct steps to perform installation and testing.

Subtask Getting to Know the Sorting Unit

Sorting unit is the last unit in the Automatic Production Line, in which the processed and assembled workpieces sent from the last unit should be sorted and workpieces of different colors from different feeding chute should be distributed. When the photoelectric sensor on the feeding end detects workpieces on the conveyor belt which are delivered from the transport station, the frequency converter is started and workpieces will be delivered into the sorting area to be sorted.

Structure of the Sorting Unit

As shown in Fig. 3-21, the sorting unit mainly consists of the delivery and sorting devices, drive device, frequency converter module, the solenoid valve set, the wiring terminal, PLC module, the base plate and so on. Delivery and sorting devices are supposed to transport the processed and assembled workpieces and sort them after being detected by the metal sensor and the fiber optic sensor. It is mainly composed of the conveyor belt, feeding chute, material pushing(sorting) cylinder, the diffusion photoelectric sensors, rotary encoders, metal sensors, fiber optic sensors and magnetic proximity sensors.

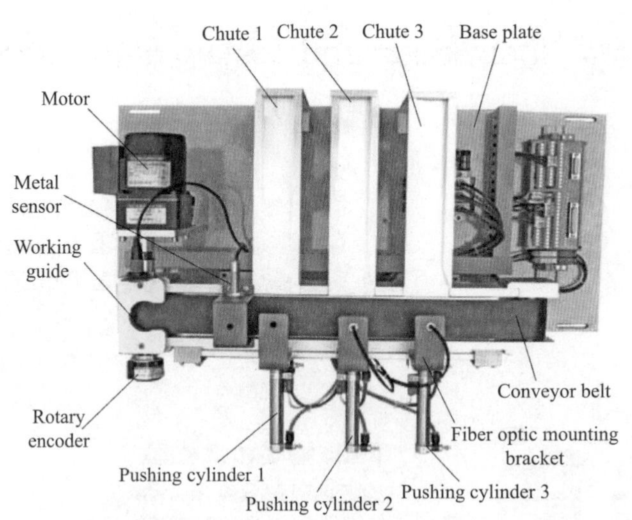

Fig. 3-21 Transporting and sorting device

The conveyor belt carries the processed workpieces from the manipulator to the sorting area. The three feeding chutes are supposed to store the processed metal, black, and white workpieces respectively.

The working process of delivery and sorting: the function of this station is to sort the assembled workpieces sent from the assembly station. When workpieces delivered from the transport station are put on the conveyor belt and are detected by the diffusion photoelectric sensor on the feeding end, signal will be passed to PLC. And accordingly, the PLC program starts the frequency converter. The conveyor belt driven by the motor starts to work and transfers workpieces to the sorting area. If workpieces entering the sorting area are metallic, then the proximity switch detecting the metal material reacts, the pushing cylinder of feeding chute 1 outputs signal and pushes the metal material to the chute 1. If workpieces entering the sorting area are white, then the fiber optic sensor detecting the white material reacts. The pushing cylinder of feeding chute 2 outputs signal and pushes the white material to the chute 2. If workpieces entering the sorting area are black, then the fiber optic sensor detecting the black material reacts, the pushing cylinder of feeding chute 3 outputs signal and pushes the black material to the chute 3; the processing of the Automatic Production Line is finished.

The pushing (sorting) cylinder installed on the opposite side of each feeding chute pushes the sorted workpieces into the corresponding chute. A magnetic proximity switch is installed on the farthest end of the three pushing (sorting) cylinders respectively. According to this signal the automatic control in the PLC can determine the current location of the sorting cylinder. When the pushing (sorting) cylinder pushes the workpiece out, the magnetic proximity switch reacts and outputs the signal of "1". On the contrary, the output signal is "0".

Precautions of Installation, Testing Delivery and the Sorting Structure

① When installing three cylinders of the sorting unit, attention should be paid to two aspects:

firstly they must be installed in the right place—workpieces should be pushed through the center of the chute, secondly they must be installed horizontally, or else they might overthrow workpieces.

② In order to make sure workpieces are pushed out from just the center of the chute accurately and smoothly, special attention should be paid to the adjustment of the three sorting cylinders' location and the stretching speed of the piston rod of each cylinder.

③ The diffusion photoelectric sensor installed in the feeding end of the conveyor belt is used to detect whether there are workpieces coming. If there are the diffusion photoelectric sensor outputs signal to PLC and then the user PLC program outputs signal to start the frequency converter, accordingly the three-phase gear motor is driven to start and carries workpieces to the sorting area. The sensitivity of the photoelectric switch should be adjusted to the extent that workpieces could be determined above the conveyor belt, because too high sensitivity could bring about interference.

④ On the top of the conveyor belt installed respectively two fiber optic sensors. The fiber optic sensor consists of a fiber optic probe and a fiber optic amplifier, with the two parts being separated. The end part of the fiber optic probe separates into two optical fibers, and they are inserted into the two fiber optic holes of the amplifier when used. The amplifier of the fiber optic proximity switch has a wide range of sensitivity. When the sensitivity of the fiber optic sensor is set low, the fiber optic probe could not receive the reflection signal from the black object that is relatively less reflective. However, the fiber optic probe could receive the reflection signal from the white object that has relatively better reflectivity. Conversely, if the sensitivity of the fiber optic sensor is set higher, the photoelectric probe could receive the reflection signal even from the black object that is less reflective. Therefore, sensitivity could be adjusted to distinguish objects in black and white to sort these two kinds of materials, thus completing the automatic sorting process.

Driving device

As shown in Fig. 3-22, the driving device employs three-phase gear motor to drag the conveyor belt to transport workpieces. Its mainly consists of the motor bracket, motor, coupling and so on.

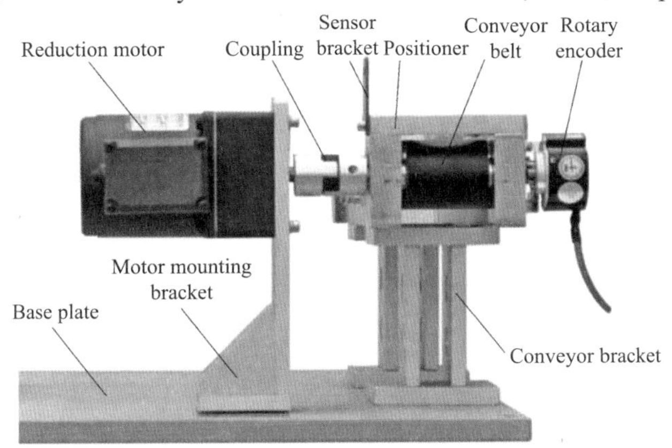

Fig. 3-22 Driving device

Motor is the main part of the driving device. The speed of the motor is controlled by the frequency converter, and the motor is to drive the conveyor belt to transport workpieces. The motor bracket is used to fix the motor. The coupling is used to connect the motor's bearing with the bearing of the drive wheel of the conveyor belt to make up a driving device.

Attention should be paid to the following points when installing and adjusting the driving device.

① The installation baseline (centerline director) of the driving device and the sliding guide rails'centerline of the transport unit should coincident.

② The motor's bearing and the bearing of drive wheel of the conveyor belt should coincident.

Summary of Knowledge and Skills

Through the training, students are supposed to get familiar with the structure of this unit, personally practicing and understanding the application of the pneumatic control technology, the sensor technology and the PLC control technology, which should be integrated organically, and experience the actual examples in specific applications of the electro-mechanical control technology.

Engineering Competence Training

Master the way of the work in engineering and develop a rigorous working style.

Task Five Installation and Testing of the Delivery Unit

The delivery unit is the most important and the most heavy-loaded part of the Automatic Production Line. In this unit, the gripping manipulator is driven to the precisely position of the material table and grip workpieces from the material table to transport it to the specified place and drops it.

The routines of the delivery unit is so great, and I will work hard!

Ask more, think more and practice more. I'm sure you will get it!

The secret of the routines lies in performing pneumatic and circuit control design according to the function of the delivery unit and completing the installation and testing following the correct procedures.

Subtask Getting to Know the Delivery Unit

The delivery unit consists of the gripping manipulator , drive components of servo motor, PLC module, the button/indicator light module, wiring terminal block and so on.

Gripping Manipulator Device

The gripping manipulator is a working unit that can realize four degree of freedom (the four-dimensional movements of lifting, stretching and withdrawing, pneumatic finger clamping/releasing and moving along the vertical shaft) movements. The device is installed on the slide board of the servo drive components. Driven by the drive components, it reciprocates in straight line and moves to the material table of other working units and finishes the function of gripping and releasing workpieces. Its structure is shown as Fig. 3-23.

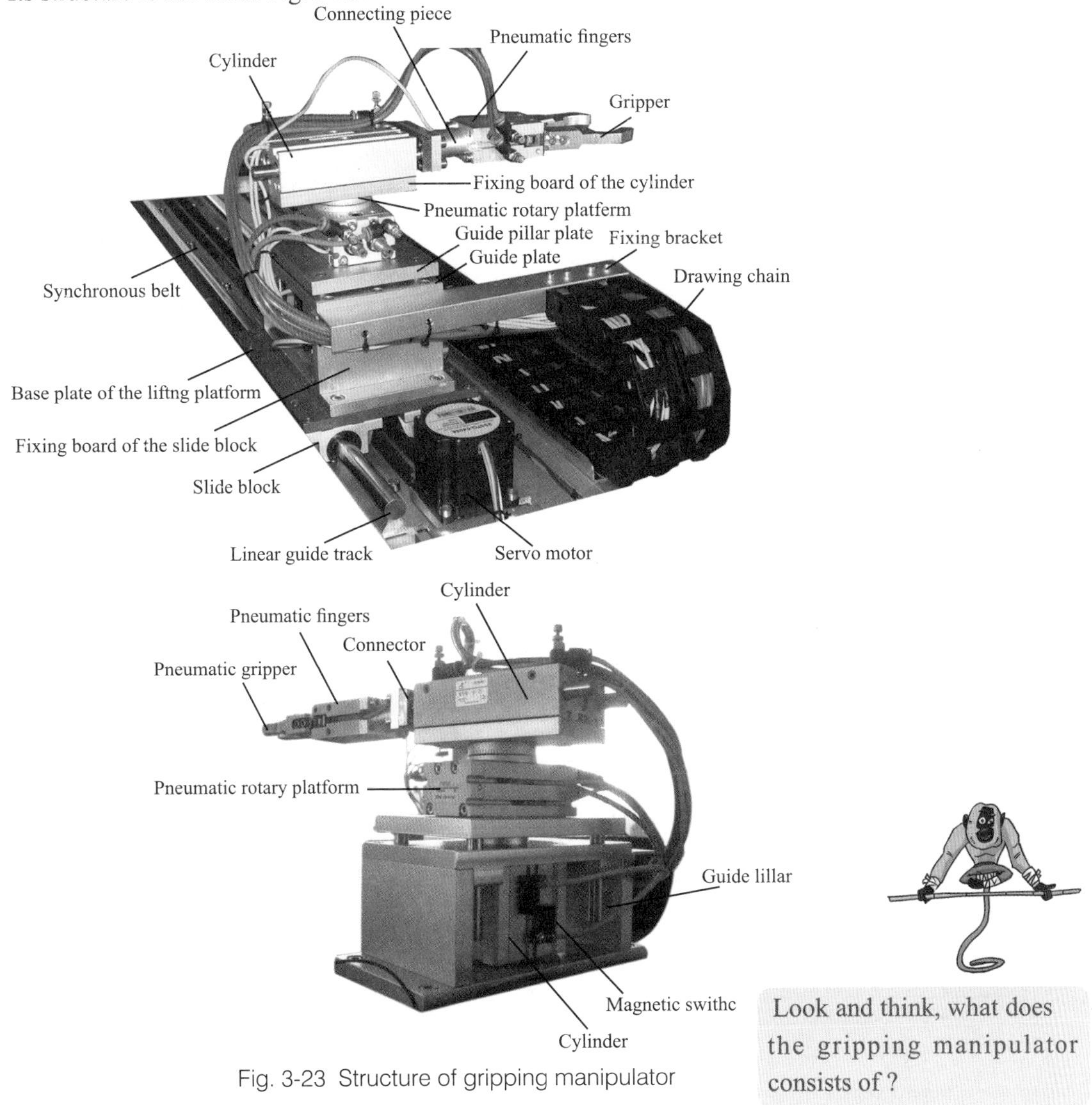

Fig. 3-23 Structure of gripping manipulator

Look and think, what does the gripping manipulator consists of ?

Specific composition is shown as follows:

① The pneumatic finger: A double acting cylinder is controlled by a two-position five-way bidirectional electrical control valve. It is used for grip and delivery the workpiece in every working station. The working principle of bidirectional electrical control valve is like the bi-stable trigger. That is, the output state is determined by the input state. Once the output state is confirmed, even there is no input, the bidirectional electrical control valve still maintains the pre-triggered state.

② Double-rod cylinder: The double acting cylinder is controlled by a two-position five-way single control valve, which is used to control the gripper's stretching and withdrawing .

③ Rotary cylinder: Double acting cylinder is controlled by a two-position five-way single control valve, which is used to control the arm's rotation of 90° forward or backward. The cylinder's rotation ranges from 0° to 180° and the adjustment is realized by two fixing buffers below the throttle valve.

④ Lifting cylinder: Double acting cylinder is controlled by a two-position five-way single control valve, which is used to control the manipulator's lifting. The running speed of cylinders above is controlled by the throttle valve's adjustment of intake flow rate in the air inlet.

Servo Drive Components

Servo drive components are used to drag gripping manipulator for reciprocating linear movement to fulfill the precise positioning. Fig. 3-24 shows the front and top view of the components, in which the gripping manipulator has been installed on the slide board of the components.

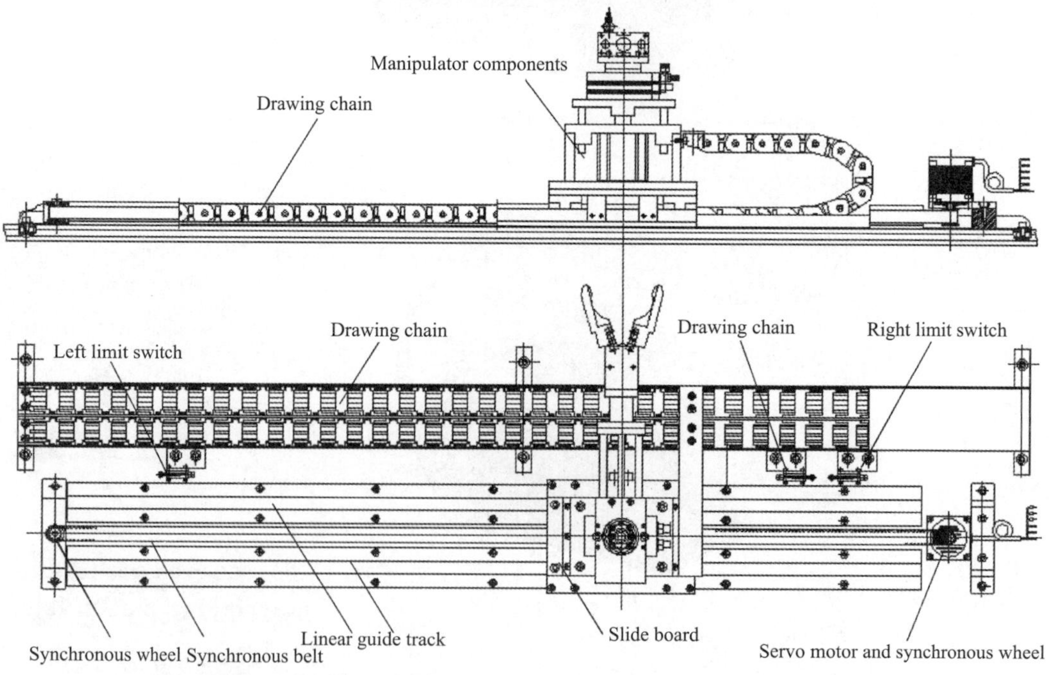

Fig. 3-24 Front and top view of the servo drive components

The drive components are composed of the servo motor, synchronous wheel, synchronous belt, linear guide track, slide board, drawing chain, the origin switch, the left limit switch and the right limit switch.

Driven by the servo driver, the servo motor brings along the slide board through the synchronous belt and the synchronous wheel to reciprocate in the linear direction along the linear guide track, and thus brings along the gripping manipulator fixed on the slide board for reciprocating linear movement.

All the air pipes and wires on the gripping manipulator are placed along the drawing chain and connected to the solenoid valve set and the connecting terminal block after entering the line groove.

The origin switch is a non-contact inductive proximity sensor. It is used to provide the signal of the starting point for the linear movement. The origin switch is installed directly on the worktable.

The left limit switch and the right limit switch are to provide the protection signal for over-travel fault. When the slide board goes over the left or right limit position, the limit switch reacts and sends the over-travel signal to the system.

The button/indicator light module is installed on the drawer-type module placement frame, with the panel layout shown in Fig. 3-25. Terminals of indicator light and buttons are all connected to the terminal board.

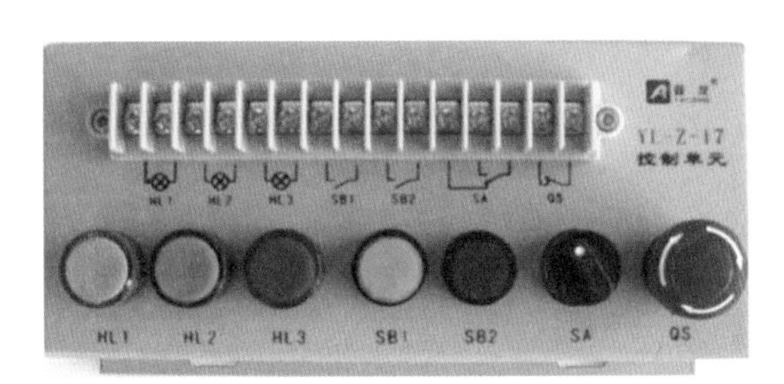

Fig. 3-25 Module of button and indicator light

Components on the module box including:

① Indicator light (24 V DC): a yellow one (HL1) , a green one (HL2) one, a red one(HL3).

② Main components: a green normally open button SB1, a red normally open button SB2, a selector switch SA (a pair of changeover contacts), an emergency stop button QS (a normally closed contact).

Summary of Knowledge and Skills

Through the training, students are supposed to get familiar with the structure of this unit, personal practicing and understanding the application of the pneumatic control technology, the sensor technology and the PLC control technology, which are integrated organically, and experiencing the actual examples in specific applications of electro-mechanical control technology.

Engineering Competence Training

Master the way of the work in engineering and develop the rigorous working style.

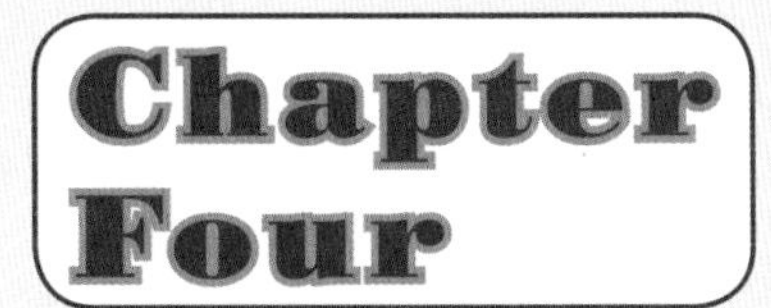

Chapter Four

Project Decision —Installation and Testing for the Automatic Production Line

Through the study of core technology learned in Chapter Two and training of installation and PLC program design of each station in Chapter Three, you are supposed to take YL-335 Automatic Production Line as an example and carry out overall installation and testing. Professional qualification integration concept is reflected during the study of this chapter. You are supposed to meet the knowledge and skill requirements of Programmable Control System Designer Professional Qualification Certificate (Level 3). YL-335B consists of feeding station, processing station, assembly station, sorting station and delivery station. The function of 5 stations is determined by work commitments of the Automatic Production Line, which reflects the great flexibility of YL-335B Automatic Production Line.

Task Objectives

1. Being able to finish the installation of Automatic Production Line within the set time;

2. Being able to implement touch screen interface set-up, web assembly and control program design of each single station according to the work commitments;

3. Being able to resolve the common problems during the installation and operation of Automatic Production Line;

4. Being able to obtain Programmable Control System Designer Professional Qualification Certificate (Level 3).

Programmable Control System Designer refers to the personnel who are engaged in PLC selection, programming, designing for the applied system, overall integration and maintenance.

Work Tasks

1. Carry out the overall design of PLC applied system and PLC ;

2. Select PLC module and determine the technical specification of products;

3. Carry out PLC programming and set-up;

4. Carry out parameter set-up of peripheral equipment and auxiliary program design;

5. Carry out control system design and overall integration, testing and maintenance.

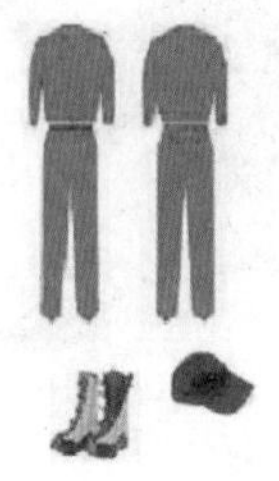

Accept Work Commitments

YL-335B Automatic Production Line consists of 5 stations: feeding, processing, assembling, sorting and delivering. Each station is equipped with one S7-200 PLC which is used to conduct the control task. PLCs connect each other with the help of PPI communication and form distributed control systems.

Work target of the Automatic Production Line:

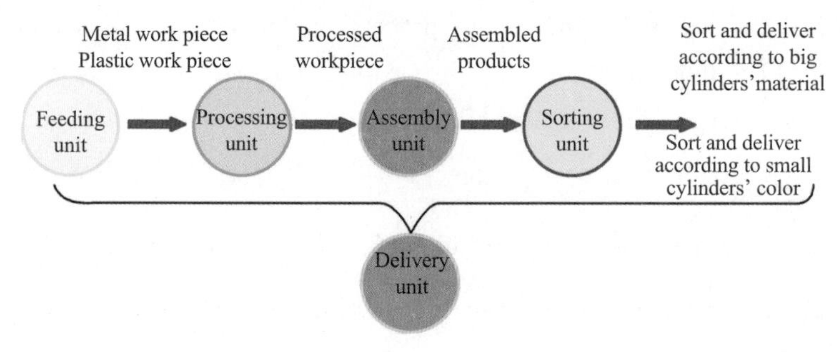

Attentions:

① The command signals of the system are provided by touch screen human-computer interface which is connected to the PLC of the delivery station.

② The working status of the whole system displays on the human-machine interface of touch screen.

③ Indicator light installed on assembly

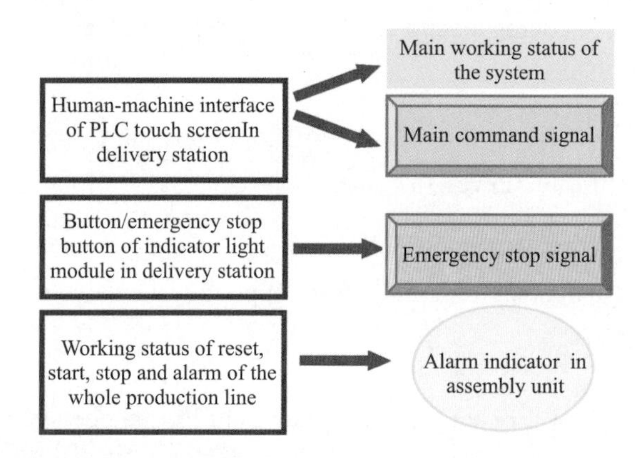

unit shows power-on reset, start-up, stop, alarm and working status of the Automatic Production Line.

④ Selection switch is available to choose working mode. You can choose single station working mode or the whole production line working mode. Under single station mode, control module of each station realizes single station control.

Be patient. You need to finish the following tasks!

Work Tasks

(1) Equipment installation

① Complete the installation of stations.

② Install all the stations on the work table.

(2) Gas circuit connection

① Design and connect gas circuit correctly.

② Use an external gas source.

(3) Electric circuit design and connection

① Design electrical control circuit of the delivery unit and connect control circuit of the the feeding unit, processing unit and assembly unit.

② Reserve I/O terminal design of the frequency converter in the sorting unit, connect main electric circuit and control circuit of the frequency convertor

③ Connect PLC communication network of each unit.

(4) Making program and doing testing

① Make PLC control program of each unit.

② Setting parameter of servo motor driver in delivery unit and setting parameter of frequency converter in sorting unit.

③ Adjust the components position of each unit, test PLC control program.

④ Touch screen is connected to PLC programming interface of the master station

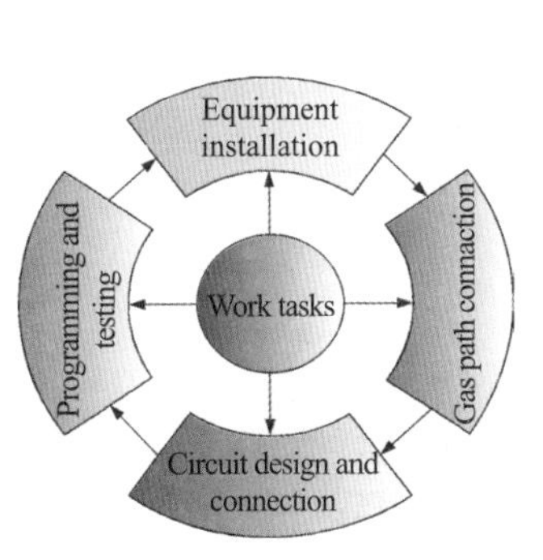

I know, we should use action-oriented teaching method taught by master.

Great! It's time for us to show what we have learned. Let's see where we should start?

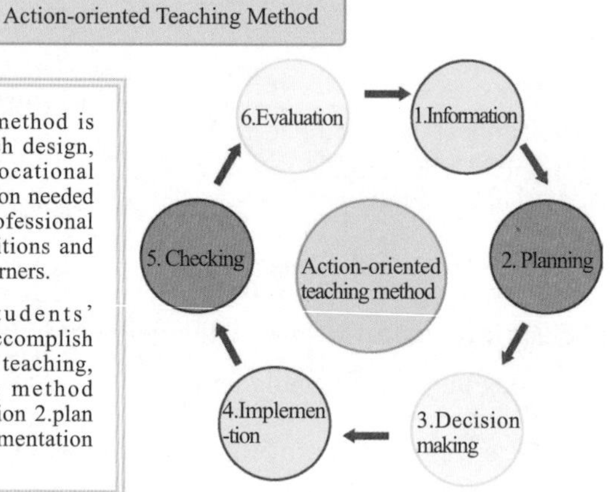

Action-oriented teaching method is the teaching activities which design, implement and evaluate vocational teaching according to the action needed when finishing a certain professional task, environment and conditions and inherent mechanism of the learners.

In order to promote students' autonomous learning and accomplish the requirements of project teaching, action-oriented teaching method includes 6 steps: 1. information 2.plan 3. decision making 4. implementation 5.checking 6. evaluation.

1. Information

What knowledge do we need to finish the tasks?

① The structure of Automatic Production Line.

② Core technology and application of the Automatic Production Line.

③ Installation and testing of each unit in the Automatic Production Line.

2. Plan and Decision

There is lots of work to do! We should make a timetable.

Attentions:

① The training time for installation is about 6-8 hours. 3 trainees should pay attention to the reasonable allocation of time and should be assigned with suitable tasks.

② According to work commitments, students can make a work plan by themselves.

Making a working plan

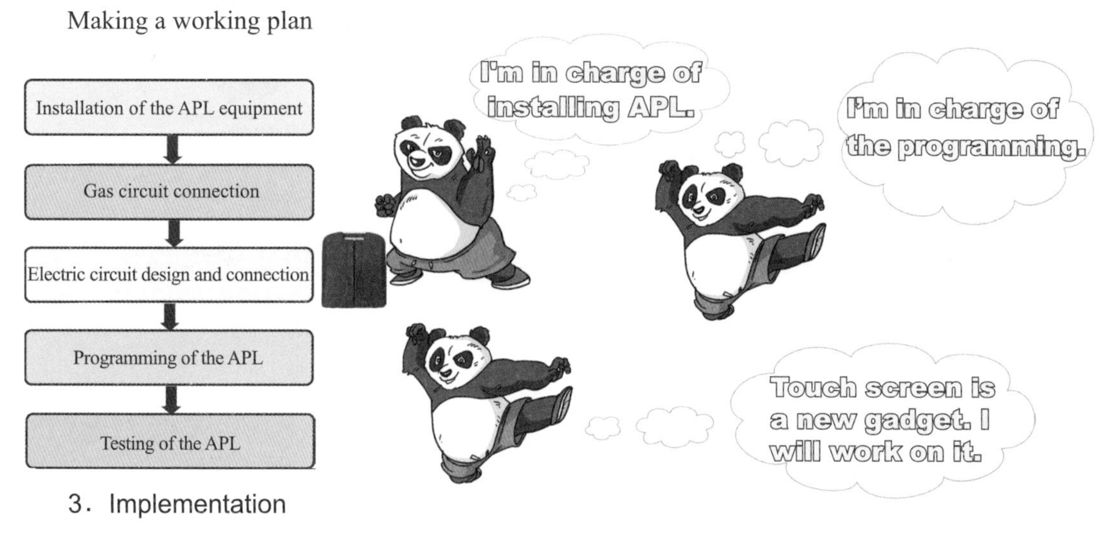

3. Implementation

 Task One Equipment Installation of YL-335B APL

1. Work Tasks

First of all, you are supposed to check whether there are all necessary components according to components' list. And then check if the quality and condition of the components meet the requirements. Then you should finish part of the assembly work for feeding, processing, assembling, sorting and delivery units. Install them on the work table of YL-335B. The working tables before and after installation See Fig. 4-1 and 4-2.Attention: Components of each unit of each unit are to be collected according to the material list.

Fig.4-1 Blank work table of APL before the installation

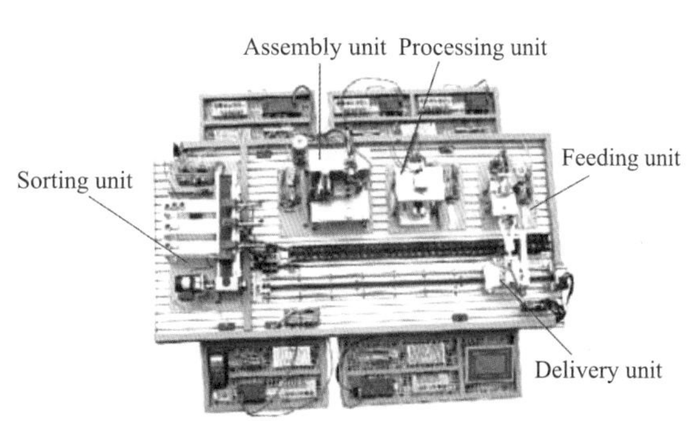

Fig.4-2 Work table of APL after the installation

2. Installation Steps

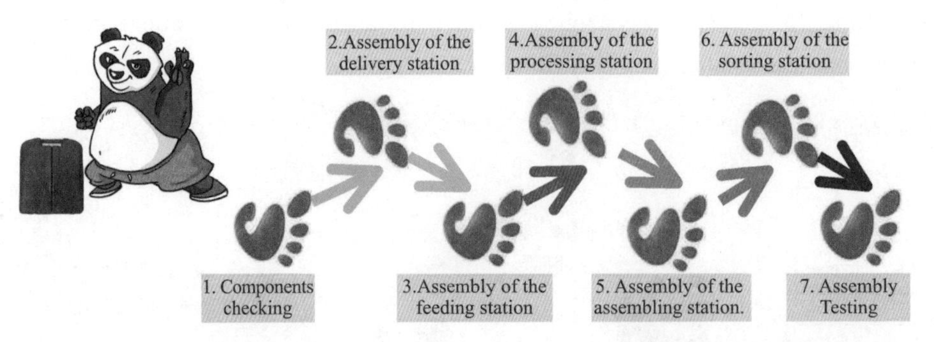

Task Two Gas Circuit Connection in YL-335B APL

Task two should be the installation of the gas circuit connection.

Work tasks

Complete the gas circuit connection of YL-335B APL based on the gas circuit knowledge learned in Chapter Two, related training of gas circuit connection of each station in Chapter Three and the control requirements of work commitments.

In task two you are supposed to design the overall diagram of gas circuit and the gas circuit diagram of each sub-station. And to connect the components of gas circuits with gas pipes of different color.

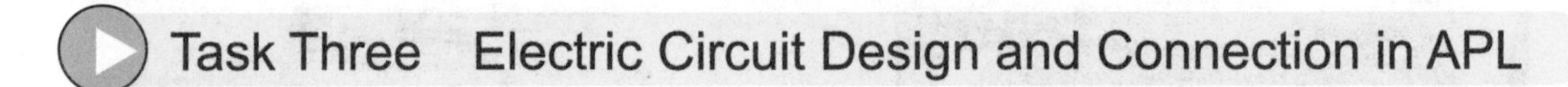

Task Three Electric Circuit Design and Connection in APL

Wow! It goes pretty well. Now let's come to task three.

Work tasks

According to the control requirements in work commitments, you are supposed to design the control circuit of the APL and connect electrical devices according to the specified PLC I /O address.

Combination with professional qualification certificates

That's true. In task three, you need to design the APL electric circuit based on the control requirements and connect the components following the electric circuit.

Hardware configuration capacity requirements of the system to obtain Programmable Control System Designer Professional Qualification Certificates (PCSDPQC). See Table 4-1.

Table 4-1 Hardware configuration capacity requirements of the system to obtain PCSDPQC

Task	Competence Requirements	Related Knowledge
Equipment Selection	1. Being able to select the PLC model, according to the capacity of input / output point, program capacity and surface sweep speed; 2. Being able to select the switching input/output unit according to the technical specification; 3. Being able to select the analog input/output unit based on technical specification and set the hardware; 4. Being able to select the suitable peripheral equipments of switching and analog unit and set the hardware; 5. Being able to calculate system power according to its configuration and select the PLC power unit and external power	1.Selection principle of PLC model; 2.Selection principle of switching input/output unit; 3.Selection principle of analog input/output unit; 4.Selection principle of PLC power unit
Reading and equipment Installation ofhardware diagram	1. Being able to read electric diagram; 2. Being able to read wiring diagram; 3. Being able to read components layout; 4. Being able to read components' site map; 5. Being able to carry out the site installation of digital quantity and analog quantity in accordance with the requirements of the drawing	1.Electric graphic symbols and drawing specifications; 2.Technical requirements of electric wiring; 3.Basic knowledge of installation and construction of electric equipment

Hah, it is getting much easier now!

Task Four Programming and Program Test

Now you are supposed to establish the network following the requirements given in work commitments (refer to the disk 4.7 work commitments) to achieve the data transmission between 5 programmable controllers; To realize each control task required in commitments based on the 5 stations program design; Master certain programming development ability.

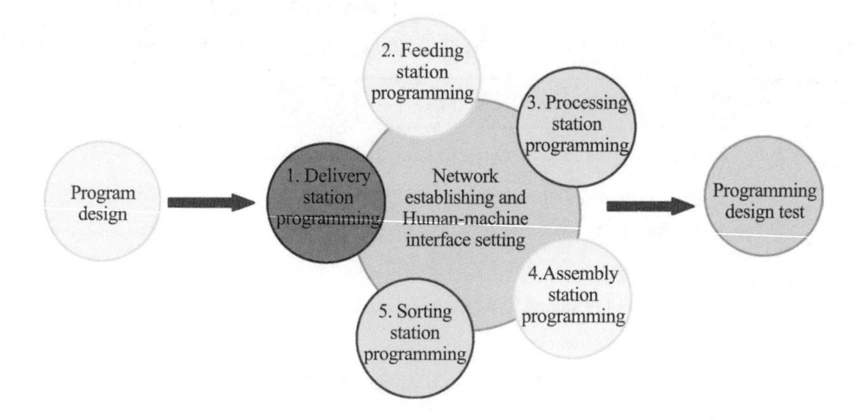

My task is 5 stations program design. That's really much to do. I've to hurry up!

I've to hurry up. After setting the human-machine interface, I have to go with installation!

Designing ability of the system to obtain Programmable Control System Designer Professional Qualification Certificates (PCSDPQC). See Table 4-2.

Combination with professional qualification certificates

Table 4-2 Designing ability of the system to obtain PCSDPQC

Task	Competence requirements	Related knowledge
Project analysis	1. Being able to analyze the process requirement of the controlled object in control system composed of digital quantity and analog quantity; 2. Being able to determine the parameters of the switching and analog quantity in single control system composed of digital quantity and analog quantity; 3. Being able to conduct statistics of switching input/output points and analog input/output points of single device control system composed of digital quantity and analog quantity. And conclude the technical specification.	1. Types of controlled object; 2. Basic knowledge of the switching; 3. Basic knowledge of analog quantity
Design of control plan	1. Being able to design block diagram of single device control system composed of digital quantity and analog quantity; 2. Being able to design the flow diagram of single device control system composed of digital quantity and analog quantity	1. Basic principle and requirements of the design of PLC control system; 2. Graphic symbols and drawing rules for flow diagram of the design of PLC control system

Summary of Knowledge and Skills

The various work stations controlled by PLC cannot realize their automatic control without PLC network communication. There are many ways of PLC network communication, and the serial communication is commonly used in industrial sites, S7-200 PLC's communication port is a physical RS-485 port. The implied communication software protocol is PPI. When a user uses reading, writing commands and guide program, he or she should make sure the setting of communications parameters should be the same between two or more communications. There is only one master station under the mode of master-slave station.

PPI is a master-slave protocol communication. The master and slave stations are in one loop net of commands. The master sends request to slave station and the slave station responses. While the slave station never sends information, just waiting for commands from master station and making response. If the PPI master mode is used in user's program, it is available to use network read / write commands in the master station to read and write information from the slave station.

Engineering Competence Training

Think about how many types of network connections are there in the PLC? How do they connect respectively? How many communication protocols are there? How do you carry on data communication? Consult S7-200 system manuals and related reference books.

Task Five　Testing and Fault Analysis in the APL

System testing requirements for obtaining Programmable Control System Designer Professional Qualification Certificates (PCSDPQC). See Table 4-3.

Combination with professional qualification certificates

Table 4-3　System testing requirements for obtaining PCSDPQC

Task	Competence requirements	Related knowledge
1. Signals testing	1. Check whether the connection of the site switching input/output signal is correct or not; 2. Check whether the connection of site analog input/output signal is correct or not; 3. Check whether the setting of analog input/output unit is correct or not	1. Ways of multi-meter use and other testing devices;testing equipment 2. Checking of site wiring;3. Testing method of analog quantity signal
2. Online testing	1. Test trapezoidal diagram and other control programs with programming tools; 2. On-line testing of the control program of single device control system composed of digital and analog quantity	1. Way of site testing of PLC control system 2. Way of testing for tools software

Hah! let me get it moving.

According to the system control flow chart, each production unit's control program in YL-335B APL has been accomplished. And they have been downloaded to PLC blocks of production unit. Each work unit in YL-335B can serve as an independent system. It can also form one distributed control system with network connection as well. In order to ensure that the program could fully realize the required function, testing should be carried out according to the different work mode. The system work mode is divided into the operation of single station and operation of whole production line. The choice of the work mode is realized through the work units button / selector switch in the indicator module, combined with the mode selection on the touch screen of human-machine interface.

Operation of the whole line → Operation of the single station	Operation of the single station → Operation of the whole line
① After completing the work cycle, the selector switch on the human-machine interface changes to the single station mode. ② The work mode of button/indicator module in each station selector switch is on the single station mode. ③ When a work unit becomes an independent system, the manual control mode is on. The command signals for equipment operation and the status signal for running are both given and showed by button/indicator module in the work unit.	① Each work station is off. Work mode of button/ indicator module in each station is on the whole line operation mode. ② The mode on the human-machine interface selector switch to the whole line operation mode, the system enters the whole line operation mode. ③ The main command signals (reset, start ,stop , etc.) of the APL system running are given through the touch screen human-machine interface. Meanwhile, human-machine interface also displays system running status information.

Attentions:

Under the mode of whole line operation, only button of delivery station / emergency stop button of indicator module is able to work. The main command of button / indicator module in other stations all becomes invalid.

 Summary of Knowledge and Skills

When the testing on the whole system is going on, at each work station, the use of single station button/indicator unit is adopted to test. First make sure that each work station is able to work, then start the operation of the whole system. If it does not work, the cylinders of each station should be checked. Whether the sensors are in the initial position or state, whether there are workpieces on the material platform of the slave station. Make sure all the PLC should be running, then put the conveyor manipulator in the middle. Press reset button, and then start the system. Take the appropriate ways to solve the problem according to different cases.

Testing on the whole system not only involves manual testing, but also includes automatic testing.

 Engineering Competence Training

Consult related information, learn pneumatic, sensors, PLC, frequency converter, stepping motor and drive modules knowledge, and get to know how to solve the breakdown problems.

Installation & Testing of Automatic Production Line

Chapter Five Project Challenging —Knowledge Development of the Automatic Production Line

The development of the APL is changing quickly. PROFIBUS, configuration, IPC, and Robot are widely used in the APL. We need to enrich ourselves with new knowledge continuously to face the new challenges.

The second competition is coming! Let's practice and face the new challenge.

Task Objectives

1.Understand the function, usage and features of PROFIBUS.

2.Understand the function, usage and features of MCGS configuration software.

3.Understand the function and features of the industrial controller.

4.Understand the function, usage and features of robot.

Task One PROFIBUS Technology

The bus of PROFIBUS is one of the many bus standards approved internationally nowadays. It has been widely used in the field of manufacturing, petroleum, metallurgy, paper making, tobacco and power. It is divided into 3 categories according to its applications: PROFIBUS-DP、PROFIBUS-PA and PROFIBUS-FMS.

What does PROFIBUS mean?

The task goes like this: cooperation based on division of work and self-study.

① Save costs of hardware and installation. Reduce hardware parts (I/O, terminal block, isolated grate) so that the installation is easy, fast, and low-cost.

② Save engineering costs. Easier configuration (only one set of tools for all equipment), easier for maintenance and repair, easier and faster system start-up.

③ More flexible. Improved functions, shorter fault time, accurate and reliable diagnosis data, reliable digital transmission technology.

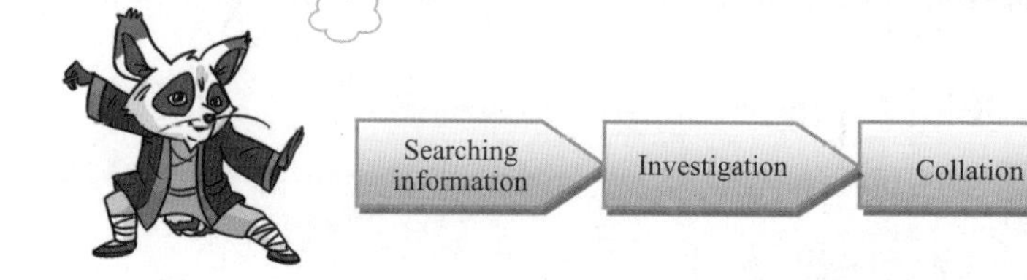

Subtask One Getting to Know PROFIBUS

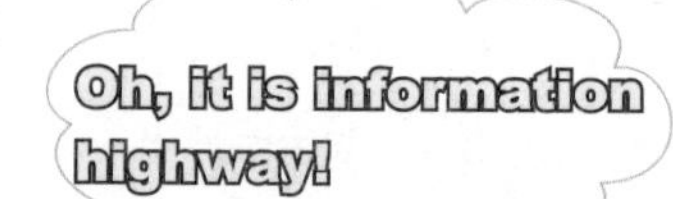

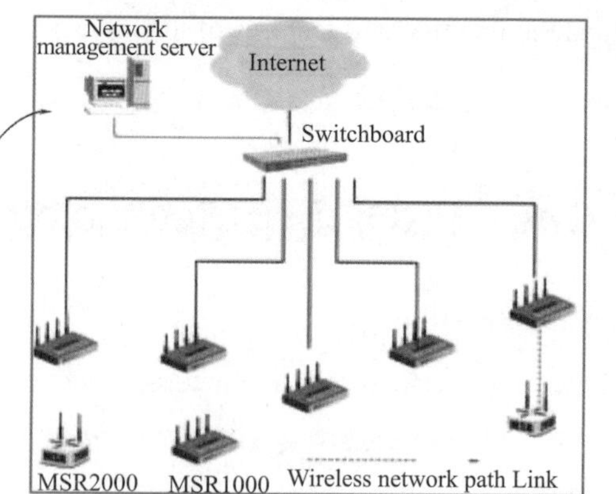

PROFIBUS Is the Leading and Open Field Bus System

Independent market survey shows that PROFIBUS is the leading technology in European market, with the highest increasing rate in the world. There are over 1,500,000 basic PROFIBUS equipment installed in 1997, with annual growth about 20% to 30%.

Information Search

PROFIBUS is an internationalized, open, site bus standard which doesn't depend on equipment producer. The transmission speed of PROFIBUS is able to be selected within the range form 9.6 kbaud to 12 Mbaud. When the bus system starts, all the devices connected to the bus should be set to the same speed. It is widely used in the automation of the manufacturing industry, automation of industrial process, automation of smart buildings, automation of transportation and power, etc. The PROFIBUS is a group of fast communications line in industrial network system. It likes the regulated highways, that is, the information highway. It has many exits. Each exit has connected a device. Fig.5-1 is the application diagram of communications network of Siemens PROFIBUS.

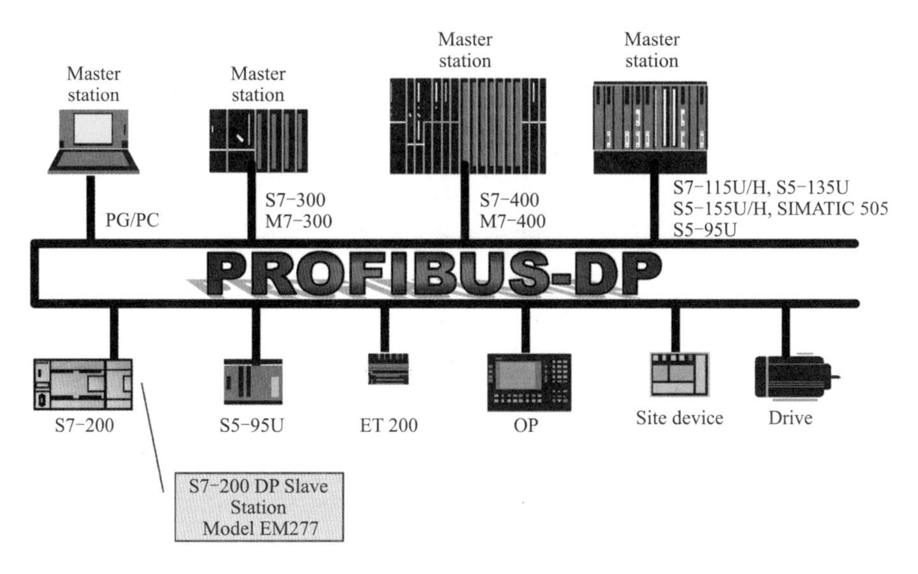

Fig. 5-1 Siemens PROFIBUS communication network

Technical Requirements of PROFIBUS Site Bus Assembly

So far, the node number of PROFIBUS has reached about 20 million in the world at present. According to an authoritative report, no fault was caused by the quality problems of the site bus device(equipment). About 95% of the problems were caused by non-regular installation. So installation is the most important basis that can guarantee the normal operation of site bus control system. Therefore, the technicians should be trained professionally so that they are qualified to carry out the installation of PROFIBUS.

Subtask Two Getting to Know the Basic Function of PROFIBUS

After obtaining some basic knowledge of PROFIBUS from the information, now let's investigate the functions and features of PROFIBUS in some enterprises.

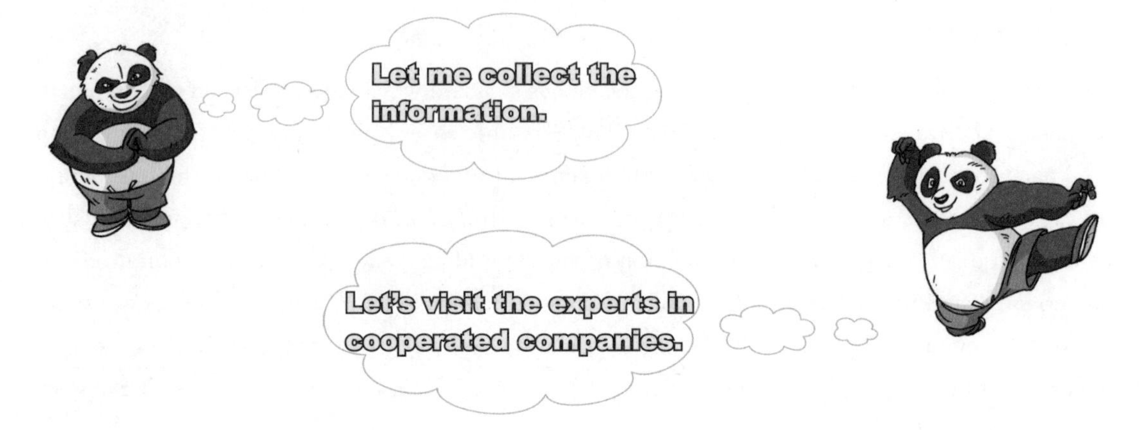

1. Structure of PROFIBUS

PROFIBUS is composed of 3 compatible parts

PROFIBUS -DP	It is mainly used for the communications between master station and slave station. That is the communication between the CPU and remote stations.
PROFIBUS -FMS	It is mainly used between master station and master station. That is the communication between the CPU and PC. It can also communicate with other producer's equipment.
PROFIBUS -PA	It is laid in the production site. Serial connection is located on the forefront devices (equipment) and apparatuses for production process monitoring.

2. Functions of PROFIBUS – DP

PROFIBUS–DP is used for the high-speed data transmission in the work site. The master station reads the information from slave station and sends the information to the slave station periodically. The circulating time of the bus must be shorter than that of master station (PLC). Besides periodically sending client's data, PROFIBUS-DP provides the non-periodicity transmission for the configuration, diagnosis and alarm treatment needed by intelligence devices. Function: There is data transmission between the DP master station circulation client and slave station circulation client. There is dynamic activation and activation availability of each DP slave station. There is configuration checking of DP slave station. Powerful diagnosis function, three level diagnosis information; Synchronization of input and output. Set the address for DP slave station through the bus. Configuration for DP master station(DPM1) through wiring. The largest input and output data from each DP slave station is 246 bytes.

Table 5-1 Main function

Transmission technology	RS-485 double strand, double cable or optical fiber cable. Baud rate is from 9.6 kbit/s to 12 Mbit/s
Synchronization	The control command allows the synchronization of input and output. Synchronization mode: synchronized output; Lock mode: synchronized input
Operation mode	Operation-clear-stop
Bus access	Command transmission among each master station. Master-slave transmission is between master station and slave station. Support single master or multi-masters system. The biggest number of stations (master-slave devices) in the bus is 126
Communications	Point to point(client data transmission) or broadcast(control command).Circulation data trans-mission of master-slave client and non-circulation data transmission of master-slaver client
Reliability	All information transmission is carried out on Hamming distance HD=4.DP slave station has the Watchdog Timer which protects the input/output storage of DP slave station. DP master station has client data transmission monitoring with viable timer
Type of device	The second type of DP Master station(DPM2)is a device that can be programmed, configured and diagnosed. The first type of DP master station (DPM1)is central PLC, e.g. PLC or PC and soon. DP slave station is with the driver, valve that has binary value or analog input and output

Table 5-2 Basic Features of PROFIBUS – DP

Speed	In a distribution system which has 32 stations, PROFIBUS-DP transmit 512 bit/s input and 512 bit/s output to all stations, and only need 1 ms at 12 Mbit/s.
Synchronization	The expanded PROFIBUS-DP diagnosis can quickly locate the fault. The diagnosis information can be transmitted through the bus and collected by the master station. The diagnosis information can be divided into three levels: Home station diagnosis operation: The common operation status of home station, like the temperature is too high, the pressure is too low Module diagnosis operation: Some specific fault of I/O module of a station Passing diagnosis operation: One single input/output fault.

Basic types of PROFIBUS-DP

Each PROFIBUS-DP system can include 3 different types of the following devices

1st level DP master station(DPM1)

The 1st level DP master station is the central controller, which can exchange information with scattered station (DP slave station) within the preset period. PLC or PC are typical DPM1.

2nd level DP master station(DPM2)

The 2nd level DP master station is a programmer, configuration devices or operation panel, which are used in DP system configuration operation, doing the system operation and monitoring.

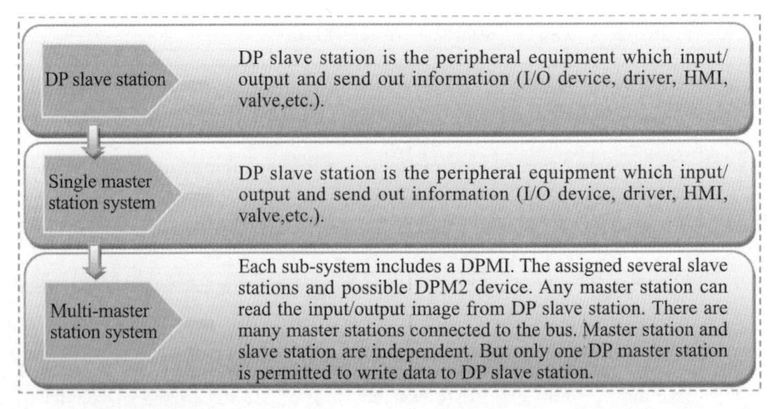

3. Regulation of Using PROFIBUS–DP

PROFIBUS–DP protocol prescribes clearly how the client data is transmitted among bus stations. But the definition of the client data is specified in the PROFIBUS regulation. It specifies how PROFIBUS—DP is applied in application fields. The regulation allows the changes and uses of different devices made by different manufactures. While the operator doesn't need to care about the differences between deices. Because the definitions that are connected with application are explained precisely in the regulation. The following is the regulation of PROFIBUS–DP. The numbers in the parentheses are file numbers.

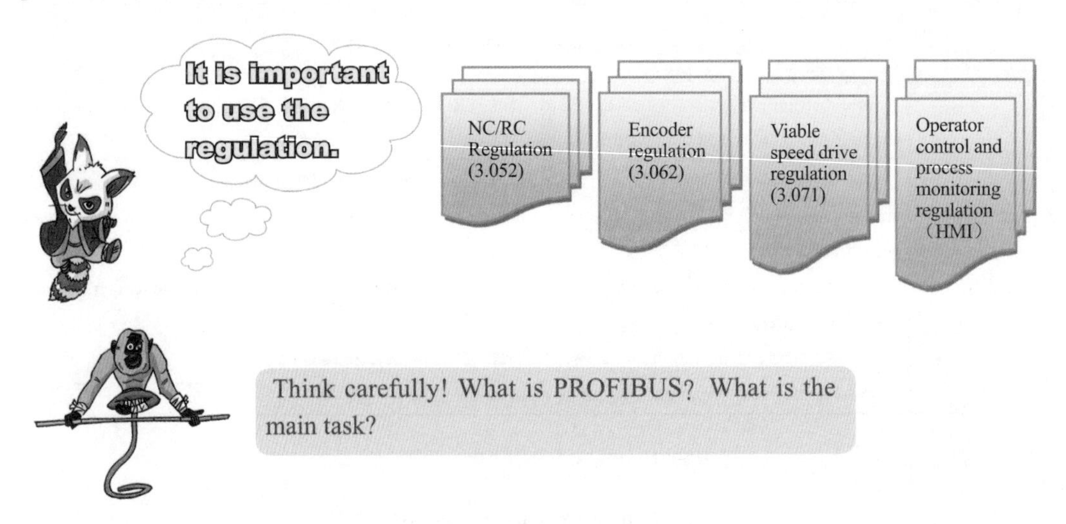

Summary of Knowledge and Skills

PROFIBUS is an internationalized, open, site bus standard which doesn't depend on equipment producer. The transmission speed of PROFIBUS can be selected within the range form 9.6 kbaud to 12 Mbaud. When the bus system starts, all the devices connected to the bus should be set to the same speed. It is widely applied in manufacturing automation, process industry automation, smart buildings automation, transportation power, etc.

Engineering Competence Training

Think for a while: What functions does PROFIBUS have? How do you classify?

 Task Two Industrial Control Configuration

Before the concept of the configuration, a task is to be completed by making programming (such as BASIC, C, FORTRAN, etc.). Making a programming is not only a hard task, a long working cycle, but also makes mistakes easily. It can't guarantee to fulfill on time. The configuration software solves this problem. The configuration software can complete the work in a few days which can be completed for several months in the past.

Configuration is so good. What is it? Follow the master's steps and practice.

Subtask One Getting to Know Configuration

In the use of industrial control software, we often mention the word "configuration". In a word, configuration is a process fulfilling a specific task with the tools and methods provided by the application software.

Taking for instance, configuration is similar to assembly.

Configuration is like building blocks!

Configuration is like assembling a computer. Prepare a variety of motherboards, chassis, power supply, CPU, indicator, hard disk, CD drives, etc. beforehand. Then assemble these components into a computer we need.

Of course configuration in the software has more space than the assembling hardware. Because it generally has more "parts" than the hardware, and each "part" is very flexible. Generally The software parts have internal properties which can be changed by changing the specifications (such as size, shape, color, etc.).

The concept of configuration first appeared in the industrial computer control, such as DCS (Distributed Control System) configuration, PLC (Programmable Logic Controller) trapezoidal diagram configuration. The software generated by man-machine interface is called industrial control configuration software.

In the field of industrial control, there is much configuration software, such as:

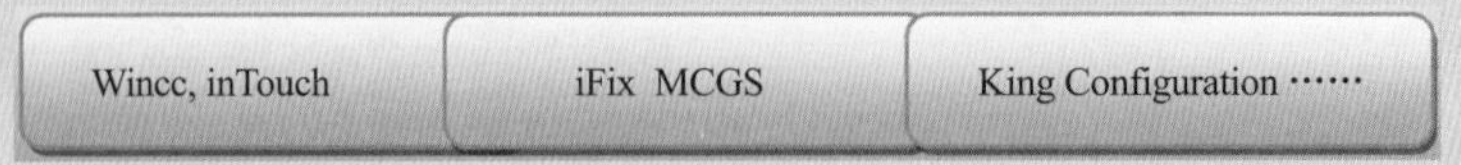

Let me summarize: configuration software is a kind of special software for data collection and process control. It is software platform and development environment at monitoring layer in the automatic control system. Using flexible configuration mode can provide users software tools which can rapidly construct monitoring functions of industrial automation control system and general level.

The applications of configuration software are very broad. It can be used for data collection, monitoring control, process control, etc. in the filed of power systems, water supply systems, petroleum, and chemical and so on.

Subtask Two Getting to Know Properties of MCGS Configuration Software

Configuration software is professional. One kind of configuration software can only be suitable for one field. The following takes MCGS configuration software (Embedded Version) as a carrier to have practice!

I go to collect the information concerning functions.

MCGS (Monitor and Control Generated System), is developed by Beijing Kunluntongtai Automation Software Technology Co., Ltd., which is a set of configuration software system based on Windows platform, used for the rapid configuration and generation of monitoring and control system. It can run on Microsoft Windows 2000/Me/NT/ operating systems and soon.

MCGS is featured by simple and flexible user interface, powerful real-time, good parallel treatment, rich and vivid multimedia screen, perfect security, powerful network functions, a variety of alarm functions, easy to control the complicated operation process.

MCGS configuration software can realize the following functions: ①Dynamic visualization control of the industrial production process; ②Production data collection and management of the production process; ③Monitoring and alarming of the production process; ④Functions of forms for reporting statistics; ⑤Web-based data upload and the corresponding control. The following online monitoring system of BMW Brilliance coating production line is to introduce these functions. Fig. 5-2 shows interface chart of online monitoring system of coating production in BMW Brilliance.

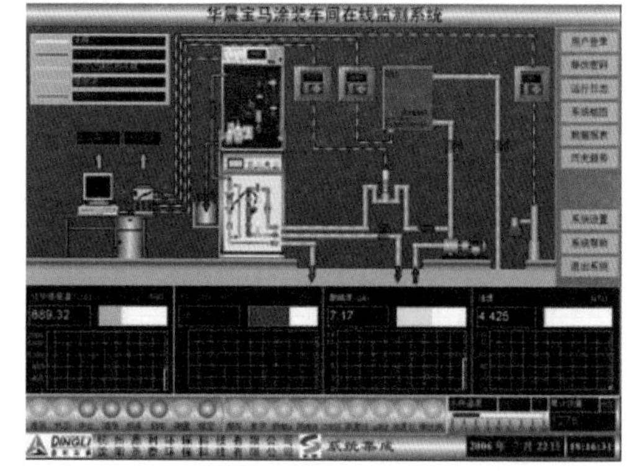

Fig. 5-2 Interface chart of online monitoring system of coating production line in BMW Brilliance

Online monitoring system of BMW Brilliance coating production lines is developed by MCGS configuration software, mainly used to help users manage the site online analysis. The software provides the following functions: ①Identity recognition; ②System flow diagram; ③Dynamic display of monitoring data; ④Operation status of indicating analyzer; ⑤Set operating parameters of the analyzer; ⑥Automatically generation of analyzer diary; ⑦Automatically generation of report forms; ⑧Inquiry report forms; ⑨Browsing historical trend curve and all local and remote operation of above features. (IE browser).

MCGS embedded version provides a good security mechanism, it can set different operating privileges for the users with different levels. In on-line monitoring system of the BMW Brilliance coating production line, the software provides 3 types of user privileges: ordinary users, leaders, super users. See Fig.5-3. Super Users enjoys special options during the testing of the system. Leaders enjoys all features except super users' privileges. Ordinary users can only browse.

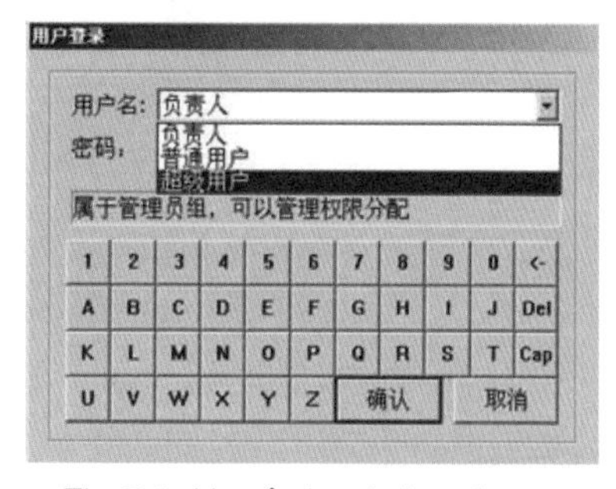

Fig.5-3 User's log-in interface

MCGS embedded version also offers users a variety of animation components. Each animation component corresponds to a specific animation function, with the size change, color change, light and dark flashes, mobile flip and other means to enhance the dynamic display effect. The status properties of definition to icons and symbols achieve animation effects. In BMW Brilliance coating production line on-line monitoring system, the software can simulate on-site system configuration through the animation. The flow chart mainly shows the user the water samples tested, the signal trend of the instruments. See Fig. 5-4.

MCGS embedded version timely offers operators the state, quality, abnormal alarm and other relevant information of the system operation in the form of the images, icons, report forms, curves and so on. In the monitoring system of BMW painting production line, the software displays the measuring value of the monitoring parameters in four visual displays: digital display, measuring range proportion display, the trend curve display, the object simulation display. See Fig.5-5.

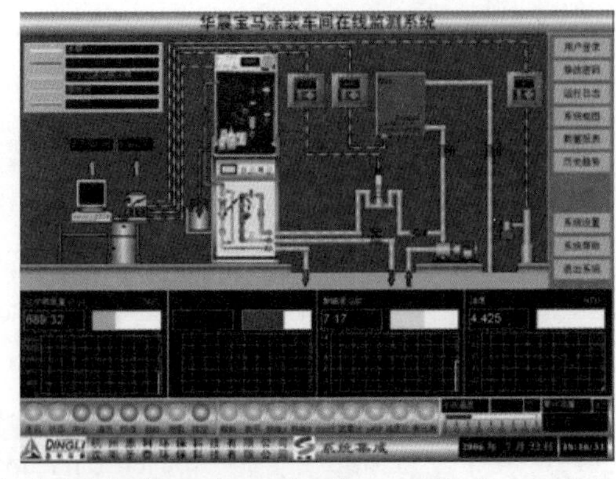

Fig.5-4 Interface chart of online monitoring system of coating production line in BMW Brilliance

Digital display is the most original, but the most effective display. It displays for the user the current value of the measurement value.

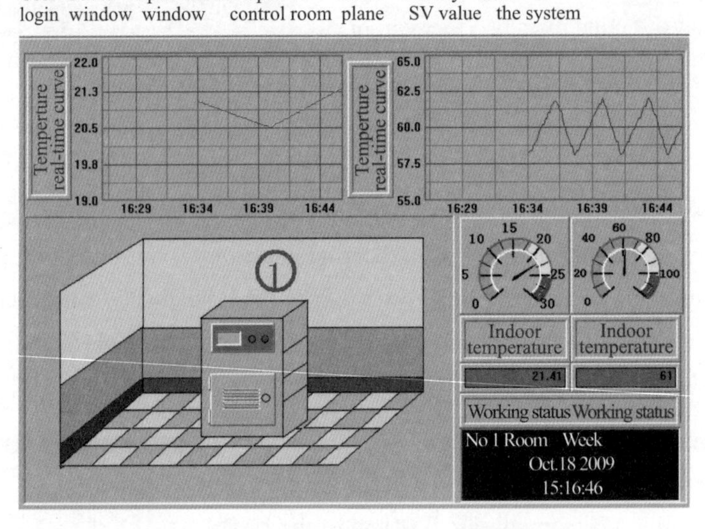

Fig.5-5 Monitoring diagram for parameters

Range proportion display shows the user the current measuring value ratio relative to the entire range in the form of the progress bar.

Trend curves display updates curve locations every minute. It can let users quickly know the status of monitoring parameters in recent period without having access to historical data.

Physical simulation display displays the measurement values based on the location of the appropriate analyzer in the flow chart, so that users view the data more visually.

MCGS is featured by real-time, good parallel processing. It takes full use of multi-task in Windows operating platform and parallel processing by priority. In BMW Brilliance coating production line on-line monitoring system, the software displays the status of the analyzer visually. See Fig. 5-6.

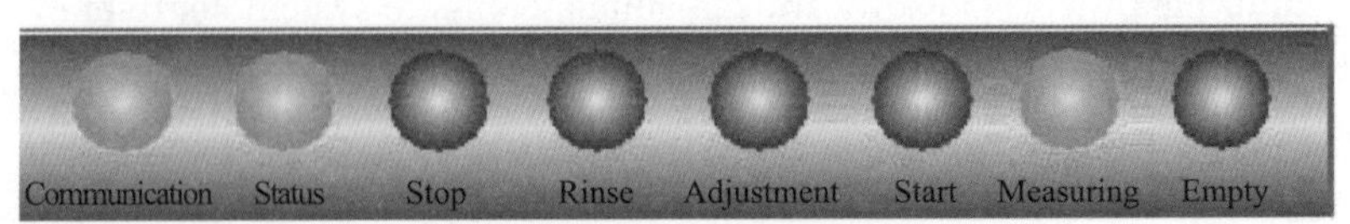

Fig. 5-6 Status monitoring chart

MCGS manages data storage by self file system or other database systems. It has high reliability. In the BMW coating production lines on-line monitoring system, in order to reflect the current water quality conditions, the software sets 10 minutes as a cycle to store the monitoring parameters automatically. You can view and print the recordings in a specified period by the inquiring function. See Fig. 5-7. You can also know the water quality conditions in a certain period through the trend curve.

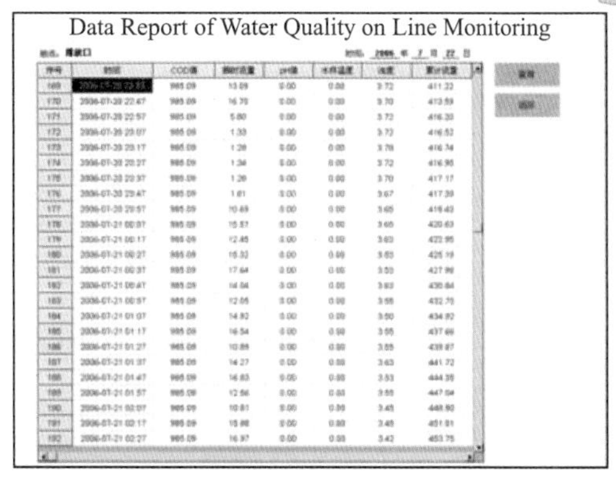

Fig.5-7 Data report chart of monitoring system

MCGS has strong function of network communication, supports serial communication, Modem serial communications, Ethernet TCP/IP communication. It not only conveniently realizes remote data transmission, but also through Web browsing, views and monitors the entire production information across the enterprise to realize the integration of devices management and enterprise management.

MCGS offers a variety of alarm modes and types. It offers users convenience to set the alarm. The system can display real-time alarm information and offer alarm data storage and response. It provides effective protection for safe, reliable industrial-site production operations.

In short, MCGS configuration software is powerful and easy to learn and use. General engineering staff can quickly grasp the design and operation of majority projects after a short period of training. MCGS configuration software can be used to avoid the complexity of the embedded version of computer software and hardware problems. And it focus on solving engineering problems in itself according to project needs and characteristics. The configuration can configure a high performance, high reliability and high professional industrial control monitoring system. Feature of MCGS embedded version configuration software shown as Table 5-3.

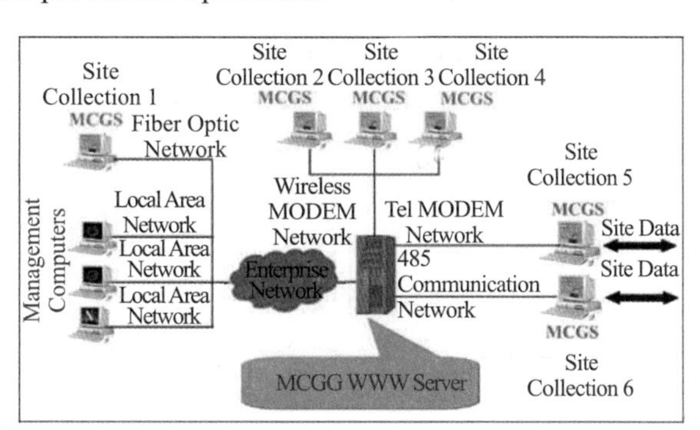

Fig.5-8 Network diagram for production monitoring

Table 5-3 Feature of MCGS embedded version configuration software

Item	Content
Little capacity	The entire system requires only 2MB storage space for minimum configuration. It is easy to use the DOC and other storage devices
High speed	High precision of the system control time can easily complete a variety of high-speed acquisition system to meet the requirements of real-time control system
Low cost	Minimum configuration of the system is 386 single board computer with main frequency 24MB, 2MB DOC, 4MB memory. It significantly reduces devices costs
Real embedded	Running on embedded real-time multi-tasks operating system
High stability	No hard disk, built-in watchdog, the power-on reset time is short. It can be long-time running constantly in a variety of harsh environment

Appendix

Item	Content
Powerful function	Provide interrupt handling, timing scan precision can reach millisecond. Provide the computer serial port, memory, ports access. And offer flexible configuration based on needs
Convenient communication	Built-in serial communications function, Ethernet communications function, Web browsing function and remote diagnostic functions can easily achieve data exchange, remote collection and Web browsing with a variety of devices
Easy to use	The configuration environment of MCGS embedded version, universal version and network version not only inherits their advantage of easy to learn, also adds flexible module operations. It forms the user control system by taking process as a unit. It makes the configuration operation of MCGS embedded version simple, visual and flexible
Support multiple Devices	Provides all commonly used drivers of hardware devices
Help to set up a complete solution	MCGS embedded version configuration environment runs on a Windows operating system with good man-machine interface. It has the same configuration environment interface as the general and network versions launched by Beijing Kunluntongtai company. It can effectively help users to build complete solutions from embedded devices, on-site monitoring stations to production monitoring information network. It is helpful projects developed by users to move smoothly on these 3 levels

Task Three Industrial Robots

Manipulator technology is an interdisciplinary comprehensive technology, which involves mechanics, machinery, electrical hydraulic technology, automatic control technology, sensor technology, computer science and other scientific fields. Currently the robots of Mitsubishi, KUKA, ABB and other companies are applied in industry. A series of industrial robots and robot system have covered all of the load levels and robot types. Such as various specifications of six-axis robot, stacking robots, gantry robots, clean room robots, stainless steel robots, high temperature resistant robots, SCARA robots, welding robots and soon. Standard robots, mounting robot and heavy-load robot can be mounted in the floor or ceiling. Its improved and flexible function, modularized industrial robots can easily and quickly be modified to adapt to other tasks. All robots work on a highly efficient and reliable computer control platform.

Subtask One Getting to Know Industrial Robots

Under ISO 8373, a robot is defined as an automatic device which may either be fixed in place or movable, realize automatic control, repeat programming, multifunction, multi application with the position of terminal operator being programmed within three or more than three degree of freedom. The degree of freedom here refers to moving or rotating axis.

Industrial manipulator is a high-tech automatic production device developed in recent decades. Industrial manipulator is an important branch of industrial robot. The feature of it is to fulfill various expected operating tasks through programming. And it absorbs the advantages of human beings and machines in both structure and properties, especially reflects human's intelligence and adaptability. The accuracy and the ability of manipulators to complete the work in a variety of environments have broad development prospects in all fields of national economy. Then let's see some common types of robots.

1. Stacking Robot

Object operations: It mainly involves the operations during storing and handling in delivery process and warehouse of the factory. The operation is to place several products on the pallet or in box when the products are delivered out of factory or stored in warehouse. If a large number of products need much manpower, it is not only difficult, but also low efficient. Stacking robot (See Fig. 5-9) could stack a large number of various products on the pallet according to the order in a short time. For example, Mitsubishi Electric palletizing robot "RV-100TH" can carry 100 kg cargos (including manipulator) at the most.

2. Sealing Operation Robot

Object operations: Install an applicator at the front end of the manipulator, for doing sealants, filler, solder coating. Sealing operation robot (See Fig.5-10) must coat the sealing parts continuously and uniformly. Therefore, the coating operating technology must be considered in teaching and programming. For example, It is necessary to take care that the waiting time of moving at the start of coating and the coating stop time in order to ensure the accuracy of track.

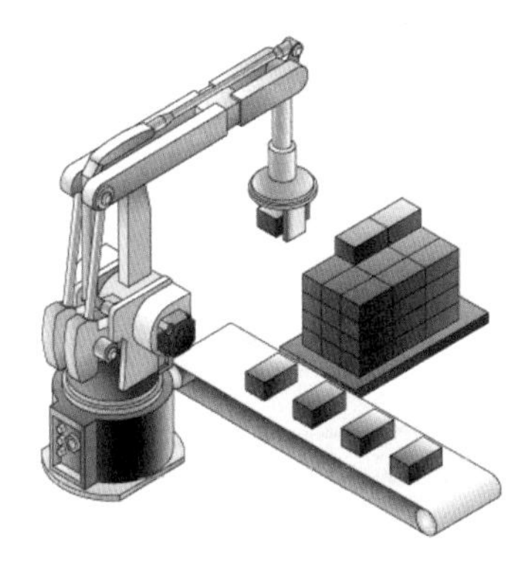

Fig.5-9 Stacking robot

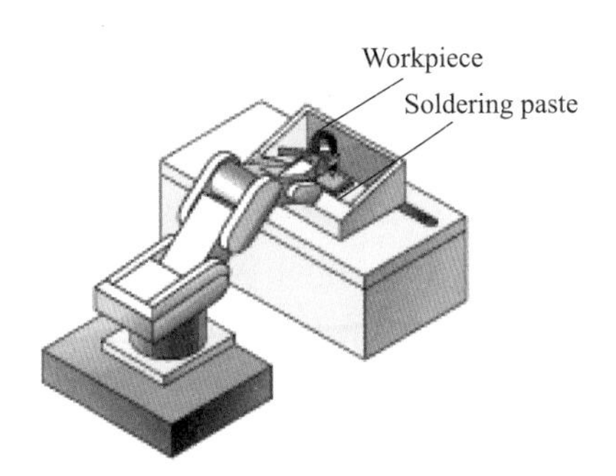

Fig. 5-10 Robot for sealing operation

3. Robot for Pouring Slot Cutting

Object operations: Cut pouring slot generated from plastic injection molding. Install the cutting tools (pliers, etc.) at the front-edge of the manipulator to do the cutting. In order to cut the slot in the complex position, robot adaptable to various postures with 5 and 6 axis vertical multi-joint, can be used. Shown as Fig. 5-11.

4. Robot for loading and Unloading Workpiece on Machine Tools

Object operations: load raw workpieces on machine tools (NC lathe) and unload the finished workpiece after processing. In the whole process of the operation, since the operation of workpiece orderly arrangement is complex, robot with 5 and 6 axis freedom is usually required, which can also withstand the dust that is generated by operating lathe are also required. Robot for workpiece loading and unloading reach this requirement. Shown as Fig. 5-12.

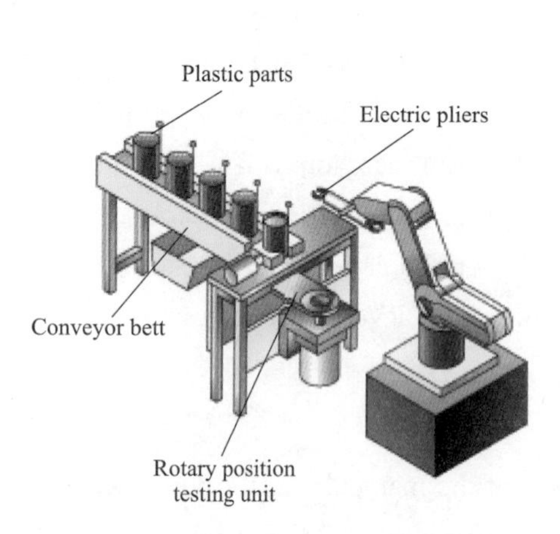

Fig. 5-11 Robot for pouring slot cutting

NC lathe
Pallet's replaceable device
Conveyor belt
Conveyor belt for unprocessed workpiece
Conveyor belt for processed workpiece

Fig. 5-12 Robot for loading and unloading workpiece

5. Robot For Clean Room

Object Operation: It is used in semiconductor and LCD manufacturing processes which needs very clean environment. Usually it works in a special "clean room". Robot for clean room is generally used in such purposes. Shown as Fig. 5-13. In short, robot is designed that do not generate dust. In case of this, all servo system is AC servo and rotating parts are sealed. In addition, the dust inside robot is emitted to the outside of the clean room through a vacuum device.

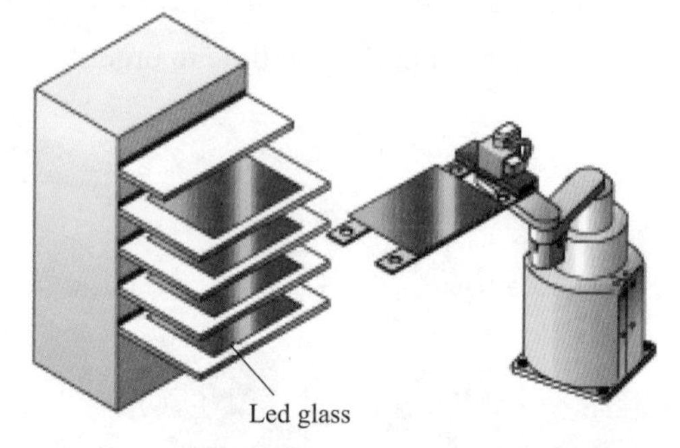

Fig. 5-13 Robot for room cleaning

Subtask Two Getting to Know the Properties of Industrial Robots

1. Functions of Industrial Manipulator

Manipulator is a multi-functions machine which can automatically position and reprogram. It has multi degrees of freedom to move objects to complete tasks in different environments.

Manipulator consists of the following five components: the actuator, driver-

transmission device, control system, intelligent system, remote diagnostic monitoring system.

The design concept of manipulator is based on human's hands and achieves the action of human beings by mechanical device. Its action and movement implements in the following four parts:

(1) Rotation of freedom;

(2) Shoulder's forward and backward movement;

(3) Elbow's upward and downward movement;

(4) Wrist (hand) movements.

Drive-transmission device and actuator are complementary. Drive system can be divided into mechanical, electrical, hydraulic, and hybrid, among which hydraulic type has the maximum operating force.

2. Classifications of Industrial Manipulator

Industrial robot can be classified as rectangular coordinates robot, cylindrical coordinates robot, polar coordinates robot, prosthetic robot, according to the structure and programming coordinate system. Industrial robots are divided into moving robot, underwater robot, cleaning robot, welding robot, surgical robot and military robot in accordance with their functional features and applications. Robotics involves in robot structure, robot vision, robot action plan, robot sensor, robot communication and robot intelligence, etc. Different robots involve different subjects. Industrial robots can be categorized according to the structure and programming coordinate system as follows:

Rectangular coordinates robot	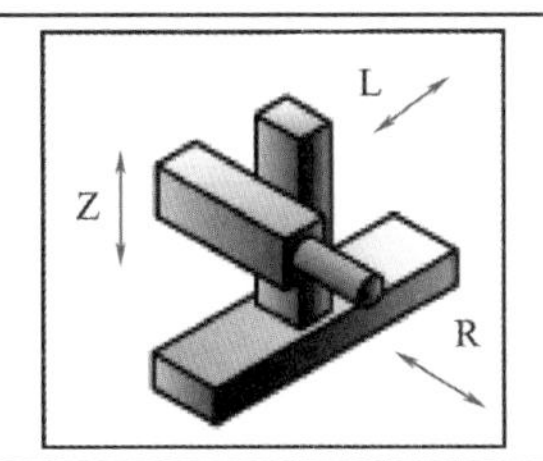 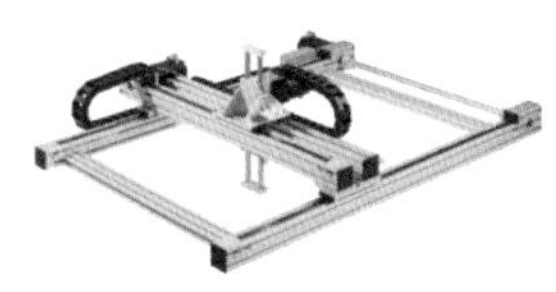
Features: rigidity, accurate positioning, easy to control; Low moving speed; Operating range is less than floor space; It is suitable to be used in loading and unloading workpieces in the assembly line, X and Y axis positioning operation, stacking operations and high precision machining.	
Cylindrical coordinate robot	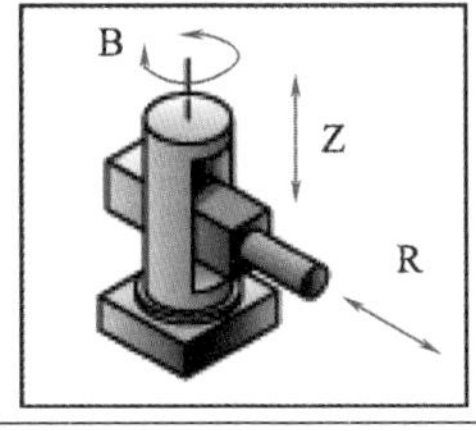 
Features: Action range is no longer limited in the front, but extended to two sides. But the upward and downward sloping movement are limited. Some complex movements like reciprocating motion are difficult to do. Rigidity, accurate positioning, easy to operate. It has rotary function, so the front edge can obtain fast linear speed. It is suitable to mechanical workpiece installation, loading and unloading operations and other packing operation.	

Polar coordinates robot		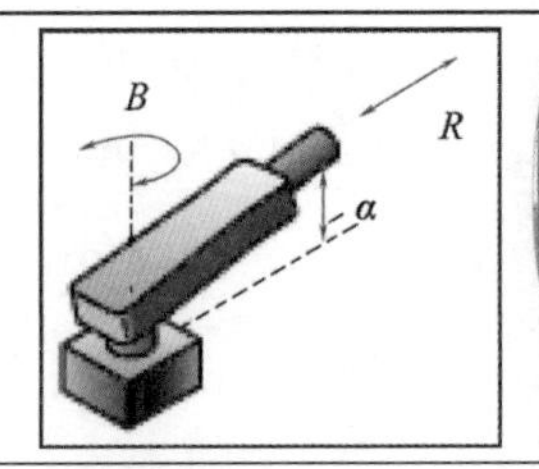 
Features: Operating space extends upward and downward. The mechanical arms can rotate up and down when it operates lower or higher than the robot body. It can do some reciprocating motion operation. The weight of the workpiece carried by the robot is less than other types of robots. It is suitable for complex space operations such as point welding, painting and surface simulation. (Currently, such kind of robot is almost not used.)		
Prosthetic robot		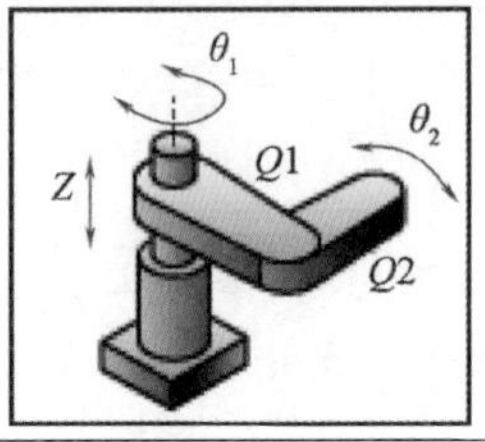 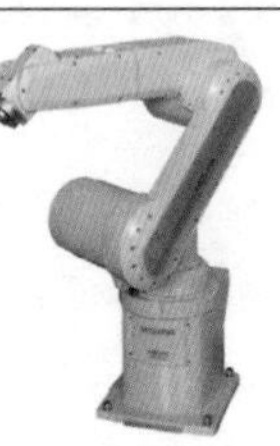
Features: Excellent function of reciprocating motion, the manipulator can move to the back side of the object to operate; It can complete complex actions; The action space is larger than floor space. Moreover, each manipulator can do circle movement. It is suitable to high speed operation. But It is less accuracy, low rigidity and low weight to be carried. Operation is more complex; It is suitable to assembly operations and complex surface follow-up operations.		

Summary of the Knowledge and Skills

Industrial robots consist of 3 basic components: main body, drive system and control system. The main body refers to the base and actuator, which includes arm, wrist and hand. Some robots have the walking part. Most industrial robots have 3 to 6 motion freedoms, of which the wrist usually has 1 to 3 motion freedoms. Drive system includes power device and transmission device which enables the actuator to produce the corresponding actions. Control system sends signals to driving system and actuator in accordance with the input program and controls them.

Robots with sense of touch, force, or simple visual can work in more complex conditions. They may become intelligent industrial robots if they have recognition function or add self-adaptation, self-learning function.

Engineering Competence Training

Look up the industrial robots information of relevant companies. Think about how to apply industrial robots in Automatic Production Line. How do you choose, install and test the robots?

Task Four　Prospects for Flexible Production Line

The characteristics of the traditional production process are single variety, large batch, special

The soft can overcome the hard, and the weak can defeat the strong!

equipment, stable process and high efficiency.

With the increasing demands for the functions and qualities of the products, products replacement cycles are getting shorter and the complexity of the products is getting increased. The way of traditional mass production has been challenged. In order to shorten the production cycle, reduce the product cost, and make small batch production eventually compete with mass production, the flexible automation systems have come into being.

Subtask One What Is Flexible Production Line?

Flexible Production Line, combining the microelectronics with the computer science and systems engineering organically, is a technically complex and highly automatic system.

Flexible Production Line is an effective means to ensure that the production adapt to changes of the market needs. It can adjust equipment combination to meet multi-machining process in accordance with the requirements. It can make the production pace of the multi-varieties and small batch similar to that of single variety and large batch. It can also greatly raise the labor productivity, lower the production costs and enhance the products qualities. Therefore, it improves the market adaptability of the enterprises. Flexibility is a comparative term of rigidness. Now, let's get to know more about its concept by the following case.

As we know: The dredger and excavators are among the main models of the construction machinery. They are quite different from each in design requirements, structure and process. According to the traditional process, both of them can' t be produced in the same production line. Hydraulic excavators have different specifications and structures as well. In the past, large, medium, and small sizes are produced by different factories. Komatsu Ltd, Japan innovated its equipment with large quantities of flexible production line in the early of 1980s so that they can organize production according to needs flexibly. As a result, the capacity of hydraulic excavator varies greatly from 0.25 m^3 to 10 m^3, (and the company has made significant growth in profits.

Want to Know the Basics of the Flexible Production Line?

Flexible Production Line is a typical application of the electro-mechanical integration technology. Now let's get to learn some basic knowledge of Flexible Production Line.

The Composition of Flexible Production Line in Mechanical Manufacturing Industry.

Since we have known the concept of Flexible Production Line, now we illustrate its composition with an example of mechanical manufacturing.

Table 5-4 Composition and Function of the Flexible Production Line

Composition	Function
Automatic processing system	It is a processing system which is based on the group technology and put the workpieces, which is in the similar dimension (not exactly the same shape), about the same weight, material and process, together on the one or several CNC machine or special lathe.
Logistics system	It is composed of a variety of transport equipment (convey or belts, and manipulator,etc.) to conduct feeding and delivery of the workpiece. It is a main part in Flexible Production Line.
Information system	It collects, handling and feedback of the information required during processing and delivering. It performs multi-level control system to machine tools and delivery equipment by computer or other control devices (hydraulic, pneumatic devices, etc.)
Software system	It is an essential part for effective management to the Flexible Production Line by computer, include designing, planning, production control, and system supervision software, etc.

The Flexible Production Line is suitable for medium and small scale batch production of about 1,000 to 100,000 pieces annually.

The Form of Flexible Production Line in Mechanical Manufacturing

The following table contains three forms of flexible manufacturing line: flexible manufacturing cell (FMC) See Fig.5-14, flexible manufacturing system (FMS) See Fig.5-15, autonomy manufacturing island (AMI), Its composition shown as the Table 5-5.

Fig.5-14 Flexible manufacturing cell (FMC)

Fig.5-15 Flexible manufacturing system (FMS)

Table 5-5 Three forms of flexible manufacturing line

Form	Composition
Flexible manufacturing cell (FMC)	FMC, consisting of 1 or 2 machining centers, has the functions of different cutters' changing and workpiece loading and unloading, delivering and storing. Besides the CNC device, there is a cell computer to manage the programs and peripheral devices. FMS fits for the small batch production and the complicated shapes of parts which needs less processing and long time processing
Flexible manufacturing system (FMS)	FMS consists of 2 or more machining centers, cleaning and testing devices. it is a better delivering and storage system for the cutters and workpiece. Apart from scheduling computer, it has process control computers, distributed numerical control terminals,etc to form the local networks with multi-stage control system
Autonomy manufacturing island (AMI)	The form of AMI is based on group technology. It base on the several CNC and general-purpose machine tools. group technology. The features of it combines process equipment with production organization, management and manufacturing. It performs the process design, CNC programs management, operation planning, real-time production scheduling,etc. It has a wide range of applications, a low investment and high flexibility

The Advantages of the Flexible Production Line

The main advantages of the Flexible Production Line, shown as table 5-6.

Table 5-6 main advantages of the flexible production line

Advantages	Definition
High utilization	The capacity of one set of machine tool increases several times than that of the single operation after being put in the flexible production line
High quality	The automatic processing system is composed of one or more machine tools. It has capable of degradation operation in case of fault. Material delivery system also has the ability to detour the malfunctioned machine tools
Stable form	The parts are loaded and unloaded once to complete the process during the processing. It has high precision and stable processing form
Flexible operation	Some of the testing, loading and maintenance can be completed during the first shaft. Unmanned normal production can be achieved during the second and third shaft. In an ideal Flexible Production Line, the monitoring system can also solve some unexpected problems, like the exchange of the worn tools, clearance of the blocked logistics and so on
Great adaptability	Tools, fixtures, and material delivery devices are adjustable. decent Layout of the system is reasonable and suitable for changing the equipments to meet the market needs

Subtask Two Getting to Know Principle of FPL Process Design

Main principle of process design of Flexible Production Line, shown as Table 5-7.

Table 5-7 Main principle of process design of Flexible Production Line

Main principle	Definition
Calculate take time of the operation(Take Time)	Operation Take Time (T.T)= effective working hours per day (s) / customer demand per day (pieces) × product module effective working hours per day (s) = working hours per day - breaks - other delays
Specify processing elements of each working post	The single machining elements are logically organized logically to form a series of operations. Since the machine cycle time (including loading, clamping, positioning, tools changing, discharging, unloading,etc.) may not be longer than that of planned cycle time, the original quota of required devices number is offered in this step. It can lessen the processing equipment if necessary. As many as parts are put on the same working post to reduce costs
Operation description	Describe the each production process and determine whether it is processing, operation, inspection, delay or storage. Processing, a valued operation, is a composite parts. The storage, inspection, and processing delay of the parts are regarded as unvalued operations. The valued and unvalued operations are assessed and recorded
Specify manual operation time of each working post	Describe the each production process and determine whether it is processing, operation, inspection, delay or storage. Processing, a valued operation, is a composite parts. The storage, inspection, and processing delay of the parts are regarded as unvalued operations. The valued and unvalued operations are assessed and recorded. Classify the described content and determine the bottleneck section of the working post for improvement
Specify the approximate numbers of workers needed	In order to determine the approximate number of workers within the planned cycle time you can use the following calculation. The approximate number of workers (MIN) = total manual operation time (s) / planned cycle time (s).As the walking and measuring time is not included in the total time of manual operation time, therefore, the approximate number of workers must be counted in the integers in calculations
Determining the work load for each worker	Study the layout of the equipment and determine the composite of the machines so that the workers can finish the assigned work in the planned cycle time. The operation time of the machines is not included. As the workers can do other work while processing, the workers walking time must be included in this step. The work should be balanced and the balance is to increase a worker's load and the idle time passes to the last worker on the last working post, who can also do some extra work. With further improved, the production capacity can be improved by labor reallocation
Complete a task combination table for each machine	Machine balanced diagram is a time sequence diagram of processing. It is used in unvalued work in recognition and machining process. Manual time bar is for value-added projects, such as clamping, unloading, auto exhausting and machining. Walking time bar is for unvalued projects, such as tool changing, positioning and fast-moving, etc
Formulate man-oriented chart	To arrange the optimal operation plan for each worker. Keep inventory in minimum to achieve the matching between production and demand. To detect the improving the labor productivity
Choose machine	The machine should be as simple as possible and has the ability of changing technology devices quickly to meet the requirements of multi-model production
Completing the scheme	Generalize each detail mentioned above to form the U-cell, L-cell or T-cell layout scheme

The design target of FPL: It is a manufacturing process with the minimum processing equipment and staff to fully satisfy the customer's needs. The design should adapt to the changes in demand, to shorten the production cycle and inventory, achieve low cost, be capacity of multi-models production and quick recovery from shutdown.

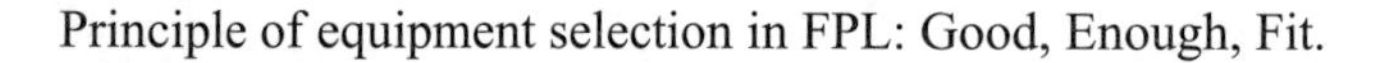

Principle of equipment selection in FPL: Good, Enough, Fit.

The principles of equipment selection for the flexible production line

The equipment selection should follow three principles: good, enough, and fit for use.

Good
Enough
Fit
Equipment should be advanced, mature, reliable, stand market test and acquire users' acceptance.
Based on the product accuracy and production program, Meeting needs is enough.
It is fit for the processing, work environment, and habits (operation habits, maintenance habits and equipment selection habits).

Configuration determines the precision, quality, efficiency, and lifespan. The quality configuration can achieve high precision, good quality, high efficiency, and long lifespan. For instance, main spindle uses electric main spindle, high-speed precision bearing, high-speed servo system, excellent CNC control system,etc. It is time to handle tools selection after choosing the machine tool. One important factor of selecting tools is to make the full use of the performance and efficiency of high-speed machine tools and avoid working under the low speed and low efficiency conditions. The overall production cost maybe much lower if high speed, high efficient and long lifespan tools are selected. The close attention should be paid to the followings:

Ah! too many things to be considered. It is impossible without experience!

1. Logistics and the Containers of the Work Station

Logistics of the production line should be simple, smooth and ensure production cycle and avoid the products damage. The logistics of the production line includes 3 kinds, shown as Fig.5-16.

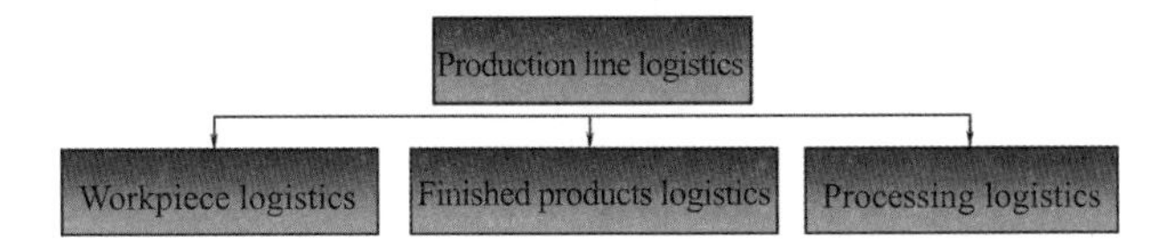

Fig.5-16 Logistics form of production line

To ensure the reliable logistics, the capacity of the working post should be considered. Strong and reliable, easy handling are fit for logistics. Intelligent rack and feeding box should be provided.

2. The Quality Control for the Whole Process.

The quality control of the production line is performed by the equipment during the processing. That is to say the good products are processed, not inspected. The quality control of production line is shown as Fig.5-7.

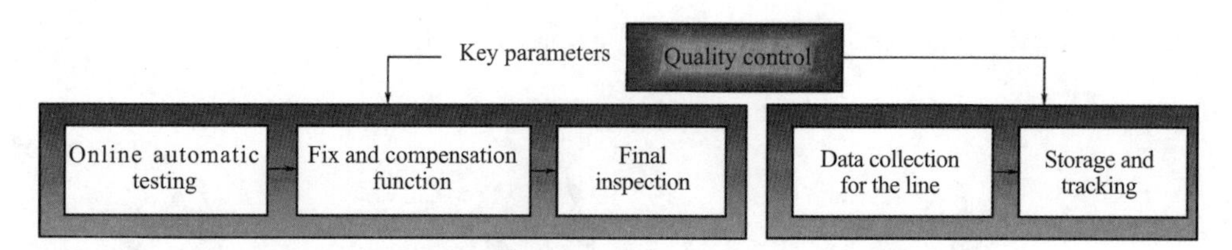

Fig.5-17 Quality control process of the production line

3. Layout of the Production Line

The layout in workshop is usually arranged vertically between assembly line and machining line. Spacing of the layout should be reasonable and proper for easy operation, logistics, maintenance and management.

The assembly line is generally ring or linear shaped layout, while the machining line is U or S shaped, or linear shaped layout. However, it is eventually decided by the size, length, capacity of the production line, the layout of the plant, etc.

Think over! What kind of technologies consist of the Flexible Manufacturing Line ? What's the applications? Can you choose devices?

 Summary of the Knowledge and Skills

Flexible production technology involves in many fields, like industry, agriculture, light industry, tertiary industry and other fields.

With the development of computer science, optics and sensor technology, the Flexible Production Line technology will make a breakthrough in low-cost, high flexibility, intelligence, environmental protection, series, modular, modern management,etc. There are two trends in its development.On the one hand, it is combined with the computer-aided design (CAD) and computer-aided manufacturing (CAM). It employs the typical process data of the original products to design different modules and constitute the different types of modular flexible system with material logistics and information logistics. On the other hand, it realizes the production processing automation for the products decision, design, production, sales and especially the computer integrated manufacturing system in management. The Flexible Production Line is only a part of the big system.

Engineering Competence Training

Think over the advantages of the Flexible Production Line, its main principles for process designing, and how to select devices for the flexible production line?

Planned Textbook by Automation Teaching Instruction Committee of Higher Vocational Education of Ministry of Education, PRC
中国教育部高职高专自动化技术类专业教学指导委员会规划教材
Achievement in Development of Teaching Resources for Items of the National Vocational Students Skills Competition in 2008
2008年中国职业院校技能大赛赛项教学资源开发成果
Achievement in National Excellent Curriculum Development"Installation & Testing of Automatic Production Line"in 2010
2010年中国国家级《自动化生产线安装与调试》精品课程建设成果

Installation & Testing of Automatic Production Line

自动化生产线安装与调试

责任编辑：祁　云　封面设计：刘　颖

地址：北京市西城区右安门西街8号
邮编：100054
网址：http://www.tdpress.com/51eds/

ISBN 978-7-113-14056-4
9 787113 140564 >

定价：38.00元（附赠CD）